Robert Henry Thurston

A Gift to
Cornell University Library
1903

A TREATISE

ON

NAVAL ARCHITECTURE.

COMPILED FROM VARIOUS STANDARD AUTHORITIES.

By Lt. Commander R. W. MEADE, U. S. N.

FOR THE USE OF THE STUDENTS OF THE U. S. NAVAL ACADEMY.

ANNAPOLIS, MD.
ROBERT F. BONSALL, Printer.

1868.

To

Vice-Admiral DAVID D. PORTER, U. S. N.,

UNDER WHOSE AUSPICES AS

SUPERINTENDENT OF THE NAVAL ACADEMY,

THE IMPORTANT STUDY OF NAVAL CONSTRUCTION

WAS REGULARLY TAKEN UP AT THAT INSTITUTION,

This Work is respectfully dedicated, by

THE COMPILER.

PREFACE

The matter contained in this book, has been mainly gathered from such standard works as those of Russell and Knowles, with some assistance from Creuze, Marrett and Peake.

A Naval Officer of fair mathematical ability, can readily make himself familiar with all the essential principles governing the design of a ship, as well as the mode of making the calculations; though to become a Naval Constructor may need the apprenticeship of the "mould loft" and ship yard.

In fact, there is no more mystery about the subject of Naval Construction than there is about the subject of Steam Enginery, and any intelligent officer may easily make himself perfectly conversant with both, while the importance of acquiring such knowledge, is self evident.

In accordance with these views, the first portion of the work, entitled "Naval Architecture," has been put together for the use of the Students at the U. S. Naval Academy.

A short treatise on "Ship-building" will be added to the next edition.

TABLE OF CONTENTS.

	Chapter.	Page.
The Science of Naval Architecture,	I.	3.
The Art of Ship-Building,	II.	4.
The Methods of Propelling Ships,	III.	7.
Different classes of Ships for Peace or War,	IV.	10.
Displacement—How to make a Ship Swim and Carry,	V.	13.
Buoyancy—Power of water to float bodies heavier than itself,	VI.	18.
Stability—Power of water to make a Ship stand upright,	VII.	22.
Powers of shoulder and under-water body,	VIII.	25.
On the proportions which make a stable or unstable ship,	IX.	30.
The method of measuring stability,	X.	33.
The powers and properties of the "shoulders,"	XI.	36.
How to give a Ship stability without great breadth of "shoulder",	XII.	38.
How to make a Ship dry and easy,	XIII.	41.
On longitudinal stability,	XIV.	47.
On the quality of weatherliness, and how to give it,	XV.	49.
How to make a Ship handy and easy to steer,	XVI.	55.
Of balance of body and balance of sail,	XVII.	58.
Of the proportion, balance, division and distribution of sail,	XVIII.	65.
Of symmetry, fashion and handiness of sail,	XIX.	76.
General conditions of the problem of Naval Architecture,	XX.	81.
How to design the lines of a Ship by the "wave system,"	XXI.	85.
On the first approximate calculation of a design,	XXII.	101.
Ships for War, including list of the iron clad ships of the United States and those of England,	XXIII.	113.
How to make a Ship drawing,	XXIV.	123.
Construction—various systems—construction of a yacht by "parabolic" system,	XXV.	127.
Mode of making calculations,	XXVI.	135.
How to set about the design of a man-of-war,	XXVII.	144.

APPENDIX.—Vocabulary of terms used in Ship-building.

CHAPTER I.

THE SCIENCE OF NAVAL ARCHITECTURE.

The Science of Naval Architecture treats of several great problems. <small>The chief problems of Naval Architecture.</small>

1st. How to make a ship swim.

2d. How to make her carry heavy weights.

3d. How to make her stand upright, when the waves or the winds try to upset her.

4th. How to make her obey the will of her Commander.

5th. How, in addition to all these, to make her go easily through the water at high speed.

Subordinate to the above, are the following:

6th. How to give a ship a given draft of water and no more; <small>Minor problems.</small> first, when she is light, and second, when she is laden.

7th. How, with the given draft of water, to prevent her oversetting when she is light, and rises high out of the water; and how to prevent her being overturned by the great burden laid upon her when she is heavily laden.

8th. How, when a heavy sea strikes on one side of the ship, to prevent it from rolling into her, without, at the same time, heeling her so far over, as to expose her to danger on the other side.

9th. How to make her bow rise to the sea, so that the waves may not roll over her deck, without, at the same time, making her rise so far as to plunge her deeply into the succeeding hollow, and make her uneasy and slow.

10th. How to make her stern of such a form that when scudding, the sea shall not break over her poop.

11th. How to make her so stiff on the water, that the pressure of the wind on her sails shall not upset her, without, at the same time, giving her so much stiffness as to endanger her masts, by the jerk of the sea.

12th. How to make her turn quickly, and in short space, in obedience to her rudder, no matter how fast she may be going; and how to make her weatherly.

13th. How, in combination with the foregoing, to make her fast before the wind, against the wind, across the wind, when she is laden, when light, when the sea is smooth, and when the sea is rough.

These are some of the undertakings with which the science of the Naval Architect must cope. They are all matters the principles of which belong to science. They are all matters of forethought and calculation, for which exact results are to be sought and ascertained, long before the ship builder can even set about his work. They form the science of Naval Architecture, as distinguished from the art of Ship Building.

CHAPTER II.

THE ART OF SHIP BUILDING.

The Art of Ship Building, consists in giving to the materials of which the ship is to consist, all the forms, dimensions, shapes, strengths, powers and movements necessary to make them fulfill and comply with the conditions resulting from the calculations of the Naval Architect.

<small>Chief problems of Ship Building.</small> 1. To make the ship swim, she must be tight and staunch every where, so as to take in no water through her seams or fastenings.

2. To make her swim so deep, and no deeper, the weights of *all* her parts, taken together, must be equal to the measure the Naval Architect has given, and which he has called her *"light displacement."* This done, it is the business of the Naval Architect, and not of the builder, to see that a given load, placed in the vessel, will not sink her beyond her given load draft.

3. To make the ship strong enough to carry her load without straining herself, is part of the art of ship building; the quantity of material put into the ship being limited by the Naval Architect, it belongs to the craft of the ship builder to select the fittest quality of material, to put it in the most effectual place, and to unite each piece in so substantial a manner, that no piece, when strained, shall part from its neighbor, but that every part shall not only do its own work, but be able to help, in need, every other part, so that all joined together, shall form one staunch whole.

4. In making the ship strong enough for the work she has to do, the builder must yet preserve, throughout the whole, such a just distribution of the weight of the parts, as that she shall not be too heavy at the bottom, nor at the top, nor at the bow, nor at the stern; but that the weights of the parts, in their places, shall so accurately correspond to the nature of the design, that there shall be a *perfect balance of weight around the exact centre*, intended by the Architect. This is necessary to be so exactly preserved, in order that the trim of the vessel, both at the bow and the stern, and her *stiffness* or power to stand upright, shall turn out to be what is meant in the plan. The best designs have failed through unnecessary weights being, in the execution of the work, placed where they did harm, instead of where they could have done good. Disposition of weight, therefore, in the hull, is an important point in practical ship building.

<small>Necessity that the Ship Builder should understand Geometry well.</small> 5. The geometry of ship building is one of the most important branches of the ship builder's art, and the exact fitting and execution of parts truly shaped, is one of the best points in which he can show his skill. The design to be executed having been put into his hand, the ship builder has first to lay it down on the mould loft floor, to its full size, next he has to divide and show on this drawing, in its full size, every part of which the ship is to consist; of each of these parts a separate and independent drawing has now to be made, and a shape or mould made from this in paper, in wood, or in iron. To this mould the ma-

terial of the ship, whether pieces of iron or wood, have to be *exactly* shaped; and these independent drawings or moulds must show every face and every dimension of each part. When it is remembered that, in every ship, consisting probably of several thousand parts, generally speaking, no two are alike, and only two, at most, resemble each other, namely, the counterpart pieces on the two opposite sides, and that every one of those pieces has probably four sides, each with a different curve from the other, and containing possibly one hundred perforations, (Iron Ship building,) which must have precise positions with reference to these curves, it will be seen that making the measurements and drawings, is a labor which must be performed with the utmost precision and intelligence, in order to have good, honest and reliable work, and requires no small amount of geometrical skill from the builder.

6. The art of the ship builder frequently extends, not only to the mere construction of the ship's hull, but also to the construction, or fitting in, of all those separate things which are not parts of the ship proper, and yet without which she cannot be sent to sea. There are parts which, if not made by the ship builder himself, must be so provided and fitted as if he had himself made them. A ship cannot be a ship without a *rudder*, without steering mechanism, without compasses and their binnacles, without anchors and their cables, without capstans and windlasses to raise and lower the anchor, without boats and davits, and the tackle to raise and lower them, without masts and yards, and standing rigging and running rigging, and sails and blocks, and all the means of placing them, and fastening, supporting and working them. There must be, also, pumps to work in case of accident, besides a large inventory of smaller things, all to be found, before a ship is a ship, or fit to go to sea. All these, for the most part, the ship builder has to find, and, while it is a matter of doubt, of opinion, of custom, or of special contract, how many, and which of them, are parts of the hull, or parts merely of the equipment of the hull, or of stores for her voyage, yet it is always in the ship builder's province to consider fully all these things, and so to arrange for them, that no unnecessary difficulties may be interposed in the way of those who have to supply and to fit them. Generally speaking, the rudder and steering-gear, the mechanism for fixing and working the anchors and cables, the masts and spars, and the means of attaching the rigging and working the sails, and boats and the ship's pumps, are reckoned part of the ship proper, and they are generally done in the ship builder's yard, while the rigging itself, the sails, the anchors, the cables, the compasses, and all the minor inventory are reckoned as "*Equipment*" only. It is part, therefore, of the craft of the ship builder to understand thoroughly, as well as to execute, that part of the equipment and the fitting which are reckoned as part of the hull.

<small>Outline of the Duties of the Ship Builder.</small>

7. But it is the finishing stroke of the ship-builder to place his vessel safely in the water. To this part of his skill belong all the traditions of launching. In this the traditional ship builder excels; for science has taught him nothing. The knowledge of launching has grown, and, with the odd variations in form, there is a wonderful unity in substance, even in different

Duties of the Ship Builder. countries. The construction of the cradle in which the infant ship is committed to the deep, of the ways which carry her from the shore into the water, of the slope on which she glides so smoothly down, even to the very mixture of soap and grease which lubricates her passage; all is known by fixed tradition, and so skilled has the long progress of practice rendered *this* finishing stroke of art, that constructors, when ordered to lengthen a ship already built, have been known to cut her in two, and to give to the after part so gentle a launch, that it stopped exactly when it had reached the point of distance from the fore part, to which the lengthening was meant to extend. So, also, when an attempt was made, as in the launch of the "Great Eastern," to bring in other than ship building skill, the result was a miserable and extravagant failure. This, therefore, is one of the points in which the ship builder cannot do better than adhere to his traditions. But along with these general principles of matured experience, there is enough variety of practice to leave the ship builder a wide choice. Some nations launch with the bow, some with the stern foremost, some broadside on. Some launch with the keel resting on the ways, the bilges clear; others launch with the bilges on the ways, and the keel clear; but in all these different modes a tolerable attention to the precepts of tradition will enable the ship builder to execute this "tour de force" with a fair certainty of success. In England, some have even ventured to carry this so far as to launch steamers with masts up, rigging fitted, and sails bent, their equipment on board, their engine and boilers fitted in them, their fires lighted and steam up; and they have left the ship-yard from the launch-ways in perfect safety, propelled by their own steam.

So ends the ship builder's duty.

CHAPTER III.

THE METHODS OF PROPELLING SHIPS.

The Naval Architect, the ship builder, and the marine engineer, represent three classes of professional skill, all of which go to the achievement of a perfect steamship. The duties of all must be successfully performed, in order that the duty of the steam-ship may also be performed successfully. It is not necessary that the three duties should be performed by three separate men, but all are essential. They may even be all performed by one man, and he may first form the design of the whole, then build the ship, and, lastly, construct the engines:* but, in theory, it is better to keep these parts separate, although, in practice, they cannot be too closely united. The problem of Steam Navigation.

Steam Navigation, or the propelling of a ship by steam, is effected by means of three great instruments. The source of the entire steam-power of a ship resides in the *boiler*, and it is the power of this boiler to produce steam, which ultimately determines the entire question of the power and speed of the ship. Boilers, therefore, are the first consideration in marine engineering. The second part, is that which applies the steam made in the boiler to the purpose of producing mechanical motion, and forms what is called the machinery, or *steam-engine*. It is by the engine that the steam is turned to use and worked; but engines accomplish their purpose better and worse; they all waste some steam in moving themselves, and not in moving the ship, some waste much steam, and do little work, others waste less steam, and do more work. It is very difficult to know how much is wasted, even by the best marine engines, and some of great reputation waste more than others of less. It is the business of the marine engineer to see that he effects the least possible waste, and gets out of his engines the maximum possible effect: but this result he can only know by taking careful measures, not merely of the work done by the steam in the engine, but also of the work given out by the engine after working itself. It is the duty of the marine engineer, thoroughly, to master all these points. The Boiler.

The Engine.

The third instrument of steam navigation, is that by which the ship is made to move. The boiler makes the steam, and the steam moves the engine merely, but not the ship. The engine has to move something which has to move the water, and which by moving the water, shall compel the ship to move. Though all three instruments move the ship, or tend to move it, it is only this last which directly touches the water, and which moves it and the ship: it is called the *motor* or *propeller*. The steam propeller is, therefore, the third instrument employed in steam navigation. The kinds of propellers are many and various; some being a single instrument, as a screw propeller; and the paddle wheel propeller, when used singly in the stern of a steamer, or in the centre of a double or twin vessel. There are also double propellers, as where two screws are used in one The Propeller.

Different varieties of Steam Propellers.

---

* This is the case in the French Navy, where the chief constructor is also the constructing engineer.

vessel, or where two paddle wheels are used on a vessel. There is also the jet propeller, (both steam and water,) the chain propeller, the stern propeller, and a host of others not now in practical use, but which it is good to know of, in order to avoid inventing them over again. There are propellers out of the water, and under water, at the sides and the bottoms of ships, at the bow as well as at the stern, and almost every place that can be named, has been selected by *somebody* for a propeller. It is the business of the practical marine engineer to devote his attention *to those modes of propelling which are in general use.* He must examine the laws which govern all steam propulsion; he must learn to measure the degree in which it can be made perfect, and the degree in which, in the nature of things, it must remain imperfect, aiming continually to get as near to perfection as possible. He must never forget, however, that it is absolutely impossible to attain this perfection; he must remember that water slips away from the propeller, and that in escaping from the propeller, it carries off power in the very act of motion. This power, said to be so lost, is called loss by slip. A scientific knowledge of the laws of propulsion enables him to judge when this slip runs to waste merely, and when, on the other hand, it is no more in quantity than is necessary to produce the propulsion of the vessel. In order to propel the vessel, the propeller must take hold of the water, and must push the water; the water will slip away from its hold, but in the very act of slipping, the propeller must dexterously lay hold of the water in just so many instants of time as to take out of it the greatest push with the least slip; *no* slip is nonsense; *much* slip is folly; as little slip as is practicable, may be fairly demanded of the competent Naval Engineer.*

Sails.

Besides propulsion by steam, there is another method of propulsion, the subject of an entirely distinct profession from that of the engineer and architect; that is, driving ships by sails, instead of steam. This is properly the business of the Naval officer. It is the Naval officer's business to know how he would best like all arrangements of the masts, sails and yards, so that he and his crew can best handle and manage them. But there is one part which the Naval Architect should do: he should thoroughly study the balance of sail, as every ship, according to the qualities of her design, will carry her sails badly, if they have not been perfectly balanced, in conformity with the peculiar properties, proportions and dimensions of each ship. It is the Naval Architect's business to provide the Naval officer with a perfect balance of sail; and it is the latter's business to know how to use it, and handle his ship properly when he has got it. Balance of sail, therefore, must be studied along with balance of body, draft of water, trim, and the other original mathematical elements of the design of a ship.

Stowage and trim, and knowledge that is requisite in a Naval Officer.

There is, finally, another point in which the professions of the Naval officer and Naval Architect touch each other very closely. This is, the trim and stowage of the ship; and the reason why the business of the sailor here touches so closely upon that of the

---

* The term Engineer is here used to denote the builder of an engine—*not* the mere "*engine driver*" of the naval service.

Architect is, that a little ignorance or folly, on the part of a Naval Officer, can neutralize and undo all that the Naval Architect and ship builder have done for the good qualities of the ship. *Stowage and trim.*

If he has not knowledge enough of the place where the centre of balance of weight of the ship is put, and does not contrive to keep it where it ought to be, but fills the ship with improper weights at improper places, he will ruin the performance, and mar the reputation, of the finest ship in the world.

But few Captains know these things thoroughly, and thereby acquire the reputation of good sailors, both for themselves and their ships. With an ignorant officer, it is impossible to know whether the ship be a good or a bad one.

Although it may not always be possible for a sailor to be also a thorough Naval Architect, inasmuch as each profession demands the study of a lifetime to learn it, yet a sailor *should know* enough of the architectural points of a ship to turn them to the best account; and it will be necessary, therefore, farther on, to investigate some of these points common to the Naval Officer, and Naval Architect.

CHAPTER IV.

DIFFERENT CLASSES OF SHIPS FOR PEACE OR WAR.

Different classes for Peace or War.
Although the principles which guide the Naval Architect in the construction of ships, and govern their behavior in the sea, are fixed and invariable, it will be the use the ship is to be put to, which must govern the Naval Architect in the application of those principles to practical use. Ships employed for purposes of commerce, for mere pleasure, or for purposes of war, must be as different in their construction as in their objects, and accordingly, the different classes of ships designed for such different uses, give rise to distinct departments of Naval Architecture.

For the purposes of war, the conditions which the Naval Architect has to fulfill, are widely different from those he has to meet in the design of a merchant vessel. The principles which guide him are the same; yet the points of practice are in some respects easier, in others more difficult. The merchant ship, in its voyages around the world, in search of freight, has to undergo all sorts of conditions of emptiness and fullness, of lightness and deepness of draft, and has to stow all sorts of cargoes, with every variety of bulk and of specific gravity; sometimes she has to carry a heavy deck load with little in her hold, and at other times, so much weight, so deep in her bottom, that it would seem to be almost impossible to re-unite two such opposite uses in the same ship.

The Ship-of-War.
The man-of-war has but one duty: to convey a known weight of guns and of men to a known place; and this kind of work, being so exactly known, ought to be infallibly and exactly done.

Necessity why results should be accurate.
That a ship-of-war, under such known conditions, should ever have a mistake made, or an inaccuracy found, in her draft of water, her stability or her speed, might seem therefore disgraceful, if it were not, unhappily, too common. The explanation which is sometimes given, is: that the people whose business it is to order these ships, are unable to settle, beforehand, what they are intended to do, and that they are generally afterwards ordered to do exactly that for which they were not originally designed.* Here it is clearly the duty of the competent Architect to refuse to construct the design of a ship until the essential elements for her construction have been authoritatively settled: for by so doing, he brings upon the profession of Naval Architecture, that disgrace which ought to fall on the shoulders of those who have the power and the ignorance to order him to make bricks without straw.

Speed.
There is one peculiarity which belongs equally to both kinds of vessels; that, whatever her load may be, she must, above all things, be fast. In commerce—time is money; in war—time is victory: and victory, the sole object of war, is entirely in the hands of the man who has the choice when and where to meet his enemy. This is an axiom, and needs no argument.

---

* This was the case with the "Double enders" during the late war—they were designed for river service, but were employed at sea, on blockade.

DIFFERENT CLASSES OF SHIPS.

To have easy movements in bad weather, is also the indispensable requisite of a good ship of both sorts; but the quality which constitutes a good sea-going vessel, may have to be given to them in different ways. *Ease of movement.*

In a merchant ship, the lading of the ship being variable, and its arrangement entirely under the disposition of the Captain and owner, the internal adjustment of weights may be so made, as to give her every variety of quality. In the ship-of-war, on the contrary, the disposition of weights being both invariable and inevitable, and fixed by the indispensable purpose of the vessel, the sea-going qualities must be given by the Naval Architect alone, in his original design; and the subsequent adjustment of the qualities of the ship, by disposition of weight, can be carried out only within narrow limits. It may happen, and it does happen, that the necessary disposition of the greatest weights of the ship-of-war, are hostile to the sea-going qualities of the vessel, and to the desire of the Naval Architect. The battery of the ship may be a great weight, acting high out of the water; and that will be a great difficulty, acting with great power against him.

It may be that he has to carry heavy loads of iron armor at great distances from those centres of his ship around which he is anxious to have the most complete repose, even at the time when the efforts of the sea are greatest to put those weights into violent motion; yet these very causes of bad qualities for the sea-going vessel, may form a specific virtue for the fighting vessel. The successful reconciliation of such antagonism, is the highest triumph of the skill of the Naval Architect, in the design of a ship-of-war.

A third condition of both kinds of vessel, differently carried out, according to the diversity of use, is what it will be necessary to call "capacity of endurance." In a merchant ship, sailing or steamship, this means ability to carry a large freight, to carry it at small cost, within an assigned time. To do this, a merchant ship should maintain her given speed with regularity, independently of weather, should do so at moderate wear and tear, in all the elements of her first cost, and should effect, at the same time, great economy in all the usable and consumable stores which form a great part of her floating equipment and provisions, and on which, in great measure, the profits or loss of a voyage depends. It is by these means that the Naval Architect will achieve reputation for his ship, profit for the owners, and character for himself. *Endurance. Endurance for a Merchantman.*

For a ship-of-war, the capacity of endurance must be of a nature somewhat different. She must certainly have the power of arriving with certainty at the place where she is wanted, independently of weather; but her sustaining power may often consist in her ability to keep herself in good fighting order for a long time, at a great distance from home, and, without exercising her greatest power, to be in a condition to do so at a moment's warning, without such exhaustion of her resources as may leave her helpless at a critical moment. This is a kind of economy of a very different nature from that of a merchant ship; but must be originally conferred on the vessel *Endurance for a Ship-of-War.*

DIFFERENT CLASSES OF SHIPS.

by the forethought of the Naval constructor, and must be studied and carried into effect by the wisdom and knowledge of the officer in command.

A knowledge of Gunnery, &c., necessary to the Architect.   There is another branch of professional knowledge and skill, without some acquaintance with which the Naval Architect cannot design a ship-of-war. A ship that cannot work and fire her guns when wanted, may have every other good point, and be worthless for want of that. The constructor must know, then, what is necessary, in order that the crew may work the guns to the greatest advantage, and thus aid in achieving victory.

Should two ships engage in a rough sea, the mere fact that the guns in one could be better handled than those in the other, in that state of the weather, might be the turning point of victory.

Ignorance of this point, therefore, on the part of the designer of the ship, would be failure, and he must have the knowledge of all the points relating to the placing and working of the guns before he begins his design—*not* as we frequently see, *after* the ship is built, and when it is too late.

But magnificent sailing men-of-war must be considered now as finally dismissed from service. The line of battle ship, fighting under canvas, is no longer a match for the little iron clad gunboat. It is probable that no such vessel will ever again enter into action. The production of the fleets of the future, is at present a race of competition, of science, and of skill, between the great Powers of the world. Who will win this race, must depend much upon the wisdom, forethought and capacity of the men who preside over the Navies of each country.

Taking this view of the subject, it becomes a matter of paramount necessity, that the young officers, who will eventually command our ships and lead our fleets, should thoroughly understand the conditions which regulate and control the designs of the steam fleets of modern warfare, and the methods used in their practical construction, and it is hoped that this knowledge may promote the advancement of the National interest, both political and mercantile.

CHAPTER V.

DISPLACEMENT.—HOW TO MAKE A SHIP SWIM AND CARRY.

Archimedes, the philosopher, is the founder of the principles of this branch of Naval Architecture. It was he who discovered the law of Displacement; or that *floating bodies* displace a weight of water exactly equal to their own weight. It is owing to this discovery that we understand the principles of flotation, and the best way to understand it, is to try *his* experiment, as follows: Take a tub 6 ft. long, 2 ft. wide, and 3 ft. deep, and fill it 2 ft. deep with water, mark a line where the water stands in the tub, then get into it, letting yourself sink in the water until nothing but a portion of your face floats above it; suppose you weigh 188 lbs., you will find the water in the tub to rise exactly 3 inches. *Archimedes,—the discoverer of the law of Displacement.*

It need hardly be said that the rising of the water in the tub was caused by your displacing it from the lower part of the tub which it previously filled, causing it to rise into the higher part. In short, by so much as the bulk of your body, by so much has the water been pushed out, displaced and raised. If you measure the exact bulk of your body immersed, and measure the exact bulk of the water raised, you will find them identical, bulk for bulk; but what is strange, though not obvious, is, that they are also equal, weight for weight; it was *this* that so astonished Archimedes. *Displacement a measure of bulk and of weight.*

If you wish to prove this, you may do it thus:—before getting into the tub, bore a hole just at the water's edge, and place buckets under the hole, then get in, and the displaced water will overflow; remain immersed long enough for the water to have flowed off through the hole, and reached the original mark. All the water you have displaced, is now in the bucket, and you (or rather that part of you which is immersed) occupy its former place; now get out, and weigh the water in the buckets, you will find it weighs 188 lbs., your own weight exactly. The principle of displacement, therefore, consists of two parts: *first*, that a body placed under water, displaces as much water as its own bulk; *secondly*, that it floats when it weighs less than the water it displaces. *Two parts to the principle of Displacement.*

This principle, although the foundation of ship building, has also a great many other useful applications. If you have anything of an awkward shape, and you want to measure its bulk—say a piece of wood, or a model of a boat—take a vessel of water large enough to hold it; place it where it may run over, and where the overflow of the water can be retained; put the substance under water, and measure the overflow. That, in gallons, or in cubic inches, is the exact bulk of the body. For rough and ill-shaped substances, we have no better way than this. Bodies, therefore, which are designed to float in the water, must be so designed, that when they are put into the water sufficiently far to swim just so much out of the water as is intended, *the part in the water shall be of the exact size necessary to displace* *Useful applications of this law.* *In order to design ships, what is necessary above all other considerations.*

the quantity of water intended, and that the body which floats shall be of the exact weight of the water it is designed to displace. In short, displaced bulk for immersed bulk, and weight for weight, the floating body and the water, whose place it occupies, must be identical.

Should an error arise, what will happen. Let us see what will happen if this be not accurately done; suppose the bulk of the body has been made too small for the weight which it is intended to carry, the vessel will sink deeper into the water than had been intended; and by sinking so much, it will displace the additional quantity of water necessary to make up the extra weight, and so, though it swims, will swim too deep. More displacement must therefore be found to meet the deficient weight; the vessel which was intended to swim light, will swim deep in the water, unless her weight be diminished, by lightening, until she return to her former intended depth; what is to be taken care of in the calculation, therefore, is, that at whatever depth it has been decided that the ship shall float in the water, or, which is the same thing, at whatever height the upper part is to float above the water, in that position the bulk of the part in the water, and the weight of the whole ship and its contents, must be so designed as to be exactly equal to the bulk of the water to be displaced by the ship, and the weight of the water to be so displaced.

Care necessary to be taken in calculating Displacement.

How many Displacements a Ship has. In a ship, however, it is necessary to do more than calculate one displacement. There are two critically important displacements to be calculated for every vessel.

"Light Displacement." Displacement when she is lying in the water ready to take in her guns or stores or cargo, or in the lightest state in which she will ever swim, that is, with a clean swept hold; this is called, technically, "Light Displacement." The other is "Load Displacement," which is calculated for the heaviest weight she will ever carry, and the deepest draft of water to which she will ever sink under a load. These are the two important drafts or depths of the ship in the water.

"Load Displacement."

To calculate these, the constructor must first ascertain the exact weight of the hull of the ship. He must include, in the weight of the hull, all the essential parts attached to and connected with that hull. He must add to that the full equipment necessary to fit her for sea-going use; but he must not include those stores (water, provisions, coals, &c.) which are to be consumed in actual service. This weight of hull and equipment for service, constitute the data on which to construct the light displacement of the ship.

The load displacement is next to be calculated. The data for this consists, first, of the light displacement, and secondly, in addition to this, of all the stores, provisions, water, coals, and consumable commodities to be used on the particular voyage or service intended, together with the cargo, freight, &c., of every kind which has to come on board.

To the "light displacement" corresponds what is called "the light draft" (or light line) of the ship. To the load displacement, "the load draft" (or load water line.) There is also the "light trim" of the ship, and "the load trim." In some languages, draft is called the "deep going" of the ship, and this phrase

HOW TO MAKE A SHIP SWIM AND CARRY.

gives the exact meaning of draft. "Trim" means difference of draft, or rather the difference between the depth of the after part of the ship under water, and that of the fore part, (commonly called "drag.") It is usual to give a ship such trim that the draft of water abaft is somewhat deeper than the draft forward. In this case she is said to be trimmed by the stern. If it were the contrary, she would be said to be trimmed by the head. This is what is meant when we say a ship is trimmed 2 ft. by the head, or 2 ft. by the stern; this difference of 2 ft. being technically called "the trim." When a vessel trims neither by the head or stern, but draws the same water forward and aft, she is said to be "on an even keel." It is usual to take a middle draft, half-way between the fore and after drafts, and to call it "the mean draft" of the ship, so that a ship which is trimmed to 21 ft. by the stern, and 19 ft. at the bow, is said to have "a mean draft" of 20 feet. In this case it is common also to call this 20 ft. "the draft of the ship," and to call the greatest draft of water (21 ft.) "the extreme draft;" in calculations of displacements, it is general to use the "mean draft."

The elements to be considered in calculating displacement, are as follows:

1. Dead weight when light.
2. Dead weight when laden.
3. Light draft of water.
4. Light trim.
5. Load draft of water.
6. Load trim.

These elements being settled, the constructor may calculate exactly the displacement of a ship of any given form, of which he may possess a design; first, for her light draft of water; second, for her load draft.

First.—For her light draft, the constructor marks off on the drawing of the ship, the exact part of the body of the vessel which will be under water when she floats light. He calls this "the immersed body" of the vessel, (light.) He then measures exactly, and calculates, geometrically, the bulk of this immersed body; this bulk will be expressed in so many cubic feet, say 18,000. He next takes the weight given for the ship and her equipment, when light, say 500 tons.

Now he knows that a ship will float at a given draft of water, when the quantity of water she displaces is of exactly the same weight as herself, and in this case the weight is given as 500 tons. The question, therefore, is:—whether the volume of water, namely, 18,000 ft., which is the bulk of the immersed body, (and which is therefore the quantity of water displaced,) will weigh more or less than 500 tons? *Light draft.*

Now, it will be found, that the bulk of 500 tons of water is just 18,000 cubic feet, and the displacement of the ship, as measured, is also 18,000 cubic feet; this, therefore, is the true light displacement.

Secondly.—For her load draft, he marks off on the drawing of the ship the exact part of the body of the vessel that will be *Load draft.*

under water when she is deeply laden. He then measures exactly, and calculates, geometrically, the bulk of that part of the vessel which was formerly out of the water, but which has now been sunk under it by the lading. Suppose this bulk to be 36,000 cubic feet. Thirty-six thousand cubic feet weigh 1000 tons; therefore, 1000 tons is the dead weight of cargo, which the ship will carry on the given load water line.

But the total load displacement of the ship consists, first, of the light displacement of 18,000 cubic feet; second, of the lading displacement of 36,000 cubic feet more; so that the total displacement of the ship, when laden, is the sum of the two, or 54,000 cubic feet. The immersed body of the ship at the load draft, has, therefore, a total displacement of 54,000 cubic feet; and the ship with her cargo floats a total weight of 1500 tons.

Calculating the weight a ship will carry at a given draft of water, is then a mere question of the measurement of the bulk of that part of the ship which will then be under water, and which is called the "immersed body." For every cubic foot of that immersion, the weight of a cubic foot of water is allowed, and thence is obtained the number of tons weight the water will support; this is called the "floating power" of the ship, and it really represents the buoyant power of the water, acting on the outside of the ship. The ship, itself, has no power to carry anything, or even to float; all it does is to exclude the water, and enclose the cargo. The ship is merely passive; the water carries both the ship and her cargo; and an iron ship will best illustrate this. Buoyancy is, therefore, the power of the water to carry a given ship. It is proportioned exactly to the bulk of the body of the ship under water, and its force is measured by the weight of the water displaced, and which is called the ship's displacement.

Buoyancy depends upon bulk only. The floating power of a ship has nothing to do with the shape of the ship, but is entirely due to its size or bulk. Practical ship builders, ignorant of the *laws* of Naval Architecture, have imagined that they could confer surprising powers of flotation, and ability to carry heavy weights, merely by giving certain *"proper"* shapes, imagined by themselves, to the immersed bodies of their ships. This delusion was common at one time, but has now passed away; yet it will take a great deal of thought to understand, thoroughly, why no possible invention of shape can give to a ship the power of greater or less buoyancy than is measured by the exact weight of water of her displacement. It is herein that the merit of the discovery by Archimedes consists; since the existence, at one time, of an opposite opinion, tends to show that the principle of flotation is by no means self-evident.

HOW TO MAKE A SHIP SWIM AND CARRY.

Standards of Displacement.

Weights.	Bulks.	Sizes.
*1 ton,	*36 cubic feet fresh water,	2 × 3 × 6 feet.
†1 "	†35 " " sea water,	2 × 2.5 × 7 "
*62.5 lbs.,	*1 " foot fresh water,	1 × 1 × 1 "
†64 lbs.,	*1 " " sea water,	1 × 1 × 1 "
‡10 lbs.,	1 gallon of fresh water,	6 × 6 × 7.69 inch.
1 lb.,	‡27.148 cubic inches fresh water,	3 × 1 × 9.216 "
1 ounce,	1.728 cubic inches " "	1 × 1 × 1.728 "
0.58 ounce,	1 cubic inch " "	1 × 1 × 1 "
2 tons,	72 cubic feet " "	6 × 6 × 2 feet.
5 tons,	180 cubic feet " "	6 × 6 × 5 "
10 tons,	360 cubic feet " "	6 × 6 × 10 "
100 tons,	3,600 cubic feet " "	6 × 12 × 50 "
200 tons,	7,200 cubic feet " "	6 × 12 × 100 "
1,000 tons,	36,000 cubic feet " "	12 × 24 × 125 "
10,000 tons,	360,000 cubic feet " "	24 × 50 × 300 "

Standards of Displacement.

* 62 5 pounds = 1-35.84 tons = 1-36 tons nearly, and 1 ton = 35.84 ft. *distilled* water.

† 64 lbs. = 1-35 tons exactly, and 1 ton = 36 cubic ft. salt water.

‡ The imperial gallon contains 10 lbs. of distilled water, at a temperature of 62.5 Fahrenheit—and also measures 277.274 cubic inches. If ordinary fresh water is taken at a lower temperature (say 40° Fahr.) as the standard, a cubic foot of fresh water will weigh exactly 1000 ounces, or 62.5 lbs. All the figures given above are correct, within a very small fraction. In round numbers, 36 cubic feet of fresh water, and 35 feet of sea water, measure 1 ton.

CHAPTER VI.

BUOYANCY.—POWER OF WATER TO FLOAT BODIES HEAVIER THAN ITSELF.

The power of water to float heavy bodies. Iron and steel are heavier than water, nevertheless out of them can be formed ships, which will not only float well above the surface, but will carry within them weights much heavier than themselves. Iron is nearly eight times heavier than water, and sinks like stone; lead is fourteen times heavier, and gold nineteen. Nevertheless gold and lead may be floated in ships of iron and steel; and structures, every portion of which would, if separate, sink to the bottom of the water, can be so combined as to float lightly on the top. The means by which this is accomplished, is a dextrous application of the forces of pressure of the water in such a manner, that the downward pressure of the weights on a ship shall be counteracted by an equal upward pressure from the water under the ship, and so the vessel be prevented from descending into it more than intended.

Power of water to give stability, and prevent upsetting. But this is not the only use to be made of the pressure of water. A ship, although supported from below, may roll over by its own weight, or may be overset by the force of the wind or the force of the waves; and so it becomes necessary to call in the aid of the force of the water, not merely to keep the ship from sinking, but to prevent it from being overset. In the first case, the water gives buoyancy only; in the second case, it is said to give stability also. In the former case, it gives *vertical* support; in the latter case, it gives *lateral* support. The two great services required of water are, therefore, first, buoyancy to support bodies much heavier than water; second, stability to be given to bodies which are unable to keep themselves in an upright position, without its aid.

Thus, from an element which is light, movable, and unstable, is to be drawn support and stability by the art of Naval construction. It is plain, therefore, that art and skill can have no sure foundation, except in a complete comprehension of the nature of water and of the laws which govern the application of its force.

The three properties of water. The first property of water, commonly called its liquidity, is its absolute indifference to shape; that is, it presses on all shapes equally. The second quality of water is the absolute proportion of its pressure to depth. The third property of water is the proportion of its pressure to the extent of the surface on which it presses, altogether regardless of the direction of that surface. The three elements, therefore, for the calculation of the mechanical force of water, are weight, depth and extent of surface.

Liquidity or fluidity of water, or its absolute indifference to shape. It is the liquidity of water which takes from it any tendency to assume fixed form in its own masses, (as frozen water or ice does,) or from exerting any force, as solid bodies do, to keep a shape in which it has been put. As a liquid, it will take the exact shape of any vessel into which it is poured, as well as the exact shape of any solid placed in, or on it. Therefore, to know how much any vessel of curious shape will hold, fill it with wa-

POWER OF WATER TO FLOAT HEAVY BODIES.

ter, and then empty its contents into some vessel of known size, the result is the exact capacity of the vessel.

Again, if you wish to know the bulk of anything of complicated form, plunge it into water, forcing the overflow of water into something that you can measure it with. The bulk of the displaced water is exactly what is occupied by the body now in water. This free flowing, easy running, and perfect fitting of water, seems to imply that it has no force, no resistance to moving, no power of effort. Could it be fancied that water had no weight, it might be fancied also without strength or resistance.

Weight.

As liquidity allows water to be parted hither and thither, and turned into any and every shape, indifferently, one must look for the source of its power to sustain, to resist, and to act, in its next quality, weight; which quality of matter is also indifferent to shape. The weight of a piece of iron, for example, cannot be altered by changing its shape. The weight of a quantity of water is the same whatever the shape of the vessel it may be put into, or whatever shape of outline may be given to it.

The measure of the weight, or given quantity of water, is as follows:

Quantity of water.	Weight.
1 Cubic inch.	250 grains = .036 lb.
12 " inches.	3000 " = .43 "
28 " "	7000 " = 1. "
1 " foot.	1000 ounces = 62.5 "
36 " feet.	1 ton.

These numbers are convenient to the purpose of the Naval Architect. In using them, however, he should remember that all water is not precisely alike in weight. The purer waters are represented by the above figures sufficiently well for all practical purposes. Salt water weighs more than river-water, and varies in different seas. Some sea-water is so heavy that 35 cubic feet will make a ton, instead of 36. Such salt water carries ships better than fresh, in the proportion of 36 to 35.

Different weight of different waters.

In calculations of ships in the sea, 35 feet may be conveniently taken as a ton, and 64 lbs. as the weight of a cubic foot.

The nature of the pressure of water is this, it will flow freely into any vessel into which it is allowed to run, and will fill it exactly. But if, in the bottom of the vessel, it find a hole or a weak place, it will rush out there, if not stopped by force. If force be applied to the hole, or the weak place, to prevent the escape of the water, this force is measured exactly by the height of the water above it, and by the size of the hole.

Pressure of water.

The next point in the nature of the pressure of water is, that under the pressure due to its depth, the water is indifferent to direction; if, at a foot deep, the pressure downwards is .43 lb. on an inch of surface, there is that pressure of .43 lb. on that inch, whether that inch lie with its face downwards or upwards, backwards or forwards, to the right or to the left, or in any degree of obliquity of direction. Pressure proportioned to depth, to extent of surface, but alike for all shapes, and for all directions, is characteristic of water pressure. The quantities given as the weights of water, enables one to measure exactly its pressure. If the water be a foot deep, and the hole a square inch, the pressure of the water outwards is measured (for fresh water) by the weight .43 lb.; at double the depth, .86 lb.; and for every foot of water an equal added weight. To stop it, requires just this weight applied the contrary way. The pressure of water trying to get out of a full vessel which confines it, is not different in kind or quantity from the pressure of water that surrounds a vessel, trying to get into it. If an opening be made under water in an empty vessel, like a diving bell or a ship, the water around it will press into it with just the same force as it would press out of a full vessel, because the water is indifferent to the direction of the pressure.

The pressure of the water into a vessel submerged in it is, therefore, .43 lbs. for each inch. At 36 feet deep, the pressure on one inch of the vessel is 15.5 lbs.

In a deep ship, loaded down in the water, the pressure of the water is greatest at the bottom: the water presses into the ship on every inch of "skin" with a force of .43 lb. for each foot deep under water. At 1 foot deep, the water presses inwards .43 lb.; at 7 feet deep 3 lbs. on the inch; at 28 feet deep, 12 lbs. on the inch; and at 36 feet deep, 15.5 lbs. to the inch. This is the measure of the force required to prevent water leaking into a ship through the seams of the sides and bottom, as well as the force that crushes the ship inwards, and requires strength in the hull of the ship to resist it.

Buoyancy. The power of water to float bodies is given by nothing more than the pressure of water under the vessel, which is pushing it upwards. To measure the buoyancy of a ship, is nothing more than to measure the pressure of the water on the whole bottom of the ship upwards. Let it be conceived that the ship has a flat level bottom and upright sides, and floats 10 ft. deep in the water: then the buoyancy and floating power of the ship, will be measured by the upward pressure of the water. At 10 feet below the water, this pressure is 625 lbs. upon each foot of water, or more than one-quarter of a ton. Reckoning the number of feet on the bottom to be, say 1000, the upward pressure of the water, or buoyancy, will enable the ship to carry 625,000 lbs.

In this calculation of buoyancy, the upward pressure of the water has been measured by the same rule as if it had been downward pressure, because it has already been shown that it is the characteristic property of water pressure, that it is proportionate to depth, and is not affected by direction. It is this universality of the pressure of water, with its indifference to direction, which makes the calculation of buoyancy so simple and easy. This principle of buoyancy, and its measurement, make it clear how bodies like iron, steel and brass, so much heavier than water, can be made to swim, even although, according to the law of displacement, they weigh much more than the same quantity of water.

Hollow metal will float if made thin enough. The art of making heavy bodies swim, consists then in this: to spread them out in a thin layer over so large a quantity of water, and at such a depth, that the pressure of the water upwards, shall be greater than the pressure of weight downwards.

A cubic foot of iron weighs 448 lbs., and would sink in water instantly. But take that mass and roll it out into a thin

POWER OF WATER TO FLOAT HEAVY BODIES. 21

plate 8 feet long, and 8 feet wide, and turn up its edges all around a foot deep; then the upward pressure of the water on the 36 feet of bottom, at the depth of one foot, will give 62.5 lbs. on each foot, or one ton of 2240 lbs., on the whole piece. The buoyancy, therefore, of the water on this extent of iron, is enough not only to float the original 448 lbs. forming the cubic foot, but also to carry a load of 1792 lbs. besides.

This example shows, in a striking manner, how a ship may not only be built of iron, which sinks by itself in water, but may be so built as not merely to carry its own weight of iron, but a burthen in addition four times greater than its own weight.

Such is the buoyancy of water:—and to carry any known weight, it is only necessary that the surface of the bottom of the ship be large enough, and placed at a sufficient depth below the water, to produce an aggregate upward pressure equal to the aggregate weights carried.

Table of pressure on the bottom of a Ship in Sea-water. — Table of Pressure.

Depth under Water.	Pressure on a Square Foot.	Pressure on a Square Inch.	Depth under Water.	Pressure on a Square Foot.	Pressure on a Square Inch.
Feet.	Pounds.	Pounds.	Feet.	Pounds.	Pounds.
1	64	.44	13	832	5.78
2	128	.89	14	896 = .4 ton	6.22
3	192	1.33	15	960	6.67
4	256	1.78	16	1024	7.11
5	320	2.22	17	1088	7.56
6	384	2.67	18	1152	8.
7	448 = .2 ton	3.11	19	1216	8.44
8	512	3.56	20	1280	8.89
9	576	4.	21	1324 = .6 ton	9.33
10	640	4.44	28	1792 = .8 ton	12.44
11	704	4.89	35	2240 = 1 ton	15.56
12	768	5.33			

CHAPTER VII.

STABILITY.—POWER OF WATER TO MAKE A SHIP STAND UPRIGHT.

Stability of floating bodies. That the most unstable of elements, water, should be required to confer stability, or give uprightness to heavy bodies raised to a great height above its surface, would appear to be an unreasonable expectation, were it not accomplished every day.

If it is merely imagined that the bottom of a ship is made the heaviest part, and the top the lightest, it would seem naturally to follow, as a first impression, that the bottom being the heaviest, would stay at the bottom, and the top being the lightest, would stay at the top. This disposition of weight is not what always, or often, in fact, takes place. A Mississippi or North river steamboat is 30 feet high out of the water, and but 3 to 6 ft. or so deep in it. The heavy weights of its machinery are generally high out of the water; its boilers are entirely above the water, reaching in some cases above the hurricane deck. Its cargo is carried high above the water, and its bottom, if not quite empty, is merely occupied by sleeping apartments. Such vessels, if supported on pivots at the height of the water, would certainly tumble over, bottom up.

Such vessels are certainly top-heavy, and pivoted *on land* would upset. By some power, nevertheless, in the water, they are kept upright and made to form huge floating castles, their chief weights high in the air.

It is therefore a duty to examine, and afterwards to understand and to measure, by what power water gives stability and uprightness to a large top-heavy, out of water structure.

The upward pressure on the bottom, carries weight, but does not give stability. It might be imagined, at first sight, that the upward pressure of the water on the bottom should help to give uprightness to the structure it upholds from below. But this idea will not stand examination: to push the bottom of a vessel upwards, may be simply trying to upset it, by trying to push it bottom up. What is wanted is, to keep the top up and the bottom down.

How, out of these contradictory elements, to elicit stability, is neither an obvious nor an easy investigation, for it is certain that the upward pressure of the water on the bottom of a ship, instead of being a cause of stability, is a powerful agent of instability; and that the greater it is in quantity, and the more effectual in power, the more it tends to upset the floating body. This is a view which must be thoroughly mastered by the Naval Architect.

Nevertheless, a perfect understanding of the way in which the power of water contributes to stability in a top-heavy, out of water structure, will give one a profound appreciation of, and admiration for, this remarkable quality of water. The way in which this unstable element gives stability to a top-heavy structure as it reels over is, *by continually transferring its action to the side to which the vessel is about to fall, and by continually giving a stronger push upwards, on the falling side, so as to counterbalance the falling weight, it keeps the vessel upright.*

HOW TO MAKE A SHIP STAND UPRIGHT.

A top-heavy ship is technically called "crank"—"a drunken ship"—and it really seems so; but by art, the force of water is made to pass from side to side, faster and further than the ship reels, and so it is managed, that though the vessel may heel over, she cannot capsize. The water then puts its strong pressure under the falling shoulder of the ship, and gives it a powerful lift. The way in which this shoulder is formed, the leverage with which the water acts, and the powerful lift which it gives at the right time and in the right way, is something which it requires much thought to conceive, skill to direct, and craft to apply with success. This portion of the ship may be called the "shoulder," to distinguish it from the bottom or "bilge" of the ship. *"Shoulders" give stability.*

It is the tendency of the bottom or bilge of the ship, to be pushed upwards by the water, and the pressure of the water is so great upwards, as to tend not only to keep it up, but to push it too much up and upset the vessel. One way of counteracting this, would be to put heavy weights of lead or iron on the bottom of the ship, so as to keep it always, in all circumstances, bottom down. But to put on the bottom of a ship useless weight, which one does not want to carry, is not merely a confession of great want of skill on the part of the Architect, but is a serious sacrifice of the usefulness of the ship.

It was the practice of a former day to make up for want of stability by great quantities of ballast; but the Naval Architect of the present day knows how to give sufficient shoulder to the ship, so as to make use of the fluidity of the water as a substitute for the "dead weight" of ballast; and its just application is a test of his skill.

By the shoulder of the ship, is meant that part of her side which is just about the water-line, which is sometimes a little out of, and sometimes a little under the water, as the ship reels about. It is sometimes called, for that reason, the part of the ship "between wind and water;" but it will be quite accurately defined, if it is said that the shoulder of a ship is that part which, being under the water when the ships heels over one way, is then left bare, out of the water, when she heels as far over the other way. *Shoulder of the Ship defined.*

Take, for example, a ship that has been standing upright, and has first leaned over on one side, until 2 feet of her skin are put into the water, and then leans over just as much on the other side, till 2 feet more of her skin are out of water—those 4 feet of her skin on each side which lie between these extreme positions, may be defined "the shoulders" of the ship. It is on them that she depends for power to sustain top-weight.

If from the body of the ship the two "shoulders" are taken, the remainder of the bottom, which never leaves the water, may be defined as the "under water body" of the ship, and this under water body is the part tending to upset her. The life of the ship is, therefore, a balanced effort, the underwater body continually tending to upset her, and the two shoulders, turn and turn about, trying to keep her upright. The one is the "upsetting" part of the ship—the other the "righting" part of the ship.— *Under water body, or under body defined.*

The effect of each of these contrary elements has to be measured, first—by the quantity of each element; second—by the more or less effectual manner in which it is applied. To make the upsetting body the largest in quantity, for the purpose of carrying useful loads, yet so to contrive it as to give it the least power for harm; to make the shoulders the smallest, yet so contrived as to have the most power for good;—that is the consummation of the art of the constructor.

The Meta-centre. In every ship it will be a question depending on her peculiar structure, how much of this righting power has been given to her; in other words, how much top weight she can carry, and how high out of the water she can carry it, without upsetting. The simplest way of putting it is, perhaps, to ask, at what height the whole weights of the ship, herself, and all she carries, might be kept without overpowering the stability, without overworking the shoulders, and without upsetting the ship?

This height is, therefore, a chief point to be calculated and known; it may be called the "upsetting point." It has been called the "Meta-centre," and if this be taken to mean the point beyond which you cannot go in raising the weights, it is a proper word enough. The limiting height of top weight is, therefore, the proper meaning of what is called the meta-centre. But the measure of this is manifestly one of the most important things to be known about a ship, and on which many of its good qualities depend.

CHAPTER VIII.

POWERS OF SHOULDER AND UNDER-WATER BODY.

To understand the functions of the shoulder of the ship, and those of the bottom, and the tendency of both to effect the stability of the ship, it will not be necessary to consider any but the simplest form which can float. For that purpose, take a square box of large size, say 36 feet wide, 27 feet high, and of indefinite length, and sink it by a weight 18 feet deep in the water—each foot of length of a box of these dimensions, will carry a ton weight, for every foot of its depth in fresh water; therefore, 18 feet of depth carries a weight of 18 tons. And supposing the box itself to weigh 6 tons per foot, the vessel would carry beside its own weight, a weight of 12 tons. Draw such a box, and across it the line of the surface of the water, which, call the "*water-line.*" Let the weight be represented in square form on top of it. (*See Plates, Fig.* 1.) Simple form of vessel, as an experimental shape

This box truly represents a ship, the weight truly representing a heavy deck load proposed to be carried by the ship. It may represent the weight of a man-of-war's battery, or the weight of an iron-cased battery, an iron-clad's turret, or any other top load.

The question is:—the ability, or inability of that ship to carry that weight, at that height out of the water? For this purpose, suppose it to lean over on either side, and then examine whether it tends to return to the upright position and stand up, or to overset and drop the weight into the sea. Draw the ship, therefore, in these two positions. (*Figs.* 2, 3, 4.) When this is done, it will be seen that there is a part of the ship which is never out of water, but keeps always under the water-line. This is the under-water body of the ship, or the upsetting part of the ship.

This under-water body is bounded by, first of all, the bottom of the ship; secondly, by the bilges, or corners of the bottom; and thirdly, by a water-line of the ship, in each of its two opposite positions. It is, therefore, pointed at the top, when it forms an equal sided triangle, the apex of which is in the water-line. Two flat surfaces, therefore, form the top of this under-water body, and the rest of it forms the bilges and bottom, or under-water skin of the ship. The part shaded, is that part which tends to upset the ship. Let the nature of this upsetting force produced by the under-water body, be examined. (*Figs.* 5 *and* 6.) Form of the under-water body.

To this end, observe that it is a symmetrical body, the right and left sides being of the same size, of the same shape, and in the original upright position of the body, exactly balancing on both sides. Its whole effect, then, may be assumed as concentrated in a point in its middle line. This point call B, or the centre of effort of the under-water body.

The buoyancy or upward pressure of this under-water body, will take place directly upward in the line Bb, and it will be seen that this is quite on one side of the centre of the vessel. It next to be noticed, in figs. 2 and 3, that the centre of the weight Its action tending to upset.

(W.) is on the opposite side of the upright line. When the ship careens over to the right, the weight also inclines to the right and downwards. When the ship careens over to the left, the weight also inclines to the left and downwards. The direction of its effect is marked by the downward line from W.

This effect is assisted by top weight. It will now be observed, that when the ship is lowered on the right side, the effect of the weight from above, is to press it downwards on that side. At the same moment the effect of the under-water body is equally bad in raising the opposite side out of the water. The ship is beset by two opposite forces, which, nevertheless, conspire in their bad effect; one sinks the right in the water, while the other lifts the left out of the water, so that with opposite means both tend to overset the ship. It is thus seen why the under-water body is the upsetting part of the ship. The larger it is, and the greater its power to carry weight, the more it will tend to overturn the weight it carries.

Counterpoise necessary. Some counteracting power must, therefore, be looked to, not merely to neutralize the overturning effect of the top weight and the upsetting force of the under-water body, but to do more than neutralize them, to give a balance of righting force, which shall constitute the stability of the vessel, or give her the power to right herself after she has been forced over.

The shoulders counteract the upsetting force. This righting force is found in the shoulders of the ship. These shoulders are formed by the two portions of the ship, on either side, which lie between wind and water, and are continually going in and out of the water on one side or on the other. When she leans over to the right, the whole of the right shoulder is put under water, and when she leans to the left, the whole of the left shoulder is put under water. In fig. 4, the shades on the shoulders distinguish them. Each shoulder consists of a wedge-formed body, with its point at the middle, and the base or heel of the wedge on the outside of the ship, the whole of the base or heel rising out of the water, and falling into it again as the ship careens. One-half the angle of this wedge is called "the angle of heel" or "the angle of inclination," and is taken as the measure *Angle of heel.* of the roll or careen of the ship. A ship, for example, is said to roll or careen 15° when the angle of this wedge is an angle of 30 degrees. It is convenient generally to speak of some fixed angle for this purpose, and 28° may therefore be assumed. For ships of war, it has been common to use 15°, as it is desirable for the sake of the guns, that the ship should *not* roll or careen *more* than this; but for merchant ships, under a press of sail, there is no harm in their careening 14° in, and 14° out of water, a total of 28°. These wedges or shoulders are sometimes called "the wings," sometimes "the solids of immersion and emersion." The term "shoulder," is considered the best and simplest term.

Point of effort of shoulder. To examine the effect of each shoulder, when it has been forced down under water, to rise again, and raise the top weight with it, one must consider the amount of its effort, the place where it may be reckoned as concentrated, and the direction of its effort. The quantity of its effort is measured by its buoyant power, or by the number of cubic feet of water it displaces. Its bulk in water, therefore, gives a measure of its buoyant or upward force. The place in which its buoyant effort takes effect, is about two-thirds (2-3) of its length from the point of the wedge O, (fig. 4.) The effect it produces may be assumed as concentrated in R, and its buoyant effect being directly upwards, tends to upright the vessel on the side which has been depressed under water.

But it should be observed, that that side of the ship on which the shoulder lies under water, is also that side on which the top weight tends to descend and overset; and it is, therefore, obvious, that the tendency of the shoulder is to help the descending weight to rise up again. The tendency of the shoulder is therefore to right the ship. *[The shoulder counteracts top weight.]*

But the vital question is, whether the shoulder has power enough to do so? In order to be effectual, it will not be sufficient that it should be capable of supporting the top weight merely. It has also to counteract the upsetting tendency on the opposite side, produced by the under-water body. *[Counteracts under-water body also.]*

It must not escape notice that the under-water body is always on the contrary side from the shoulder, always tending to upset the ship on that side from below, so that unless the shoulder be more powerful, and act more energetically than the under-water body, the ship will infallibly upset. The shoulder, therefore, has these two tasks at once. It must be strong enough to neutralize the upsetting force of the under body, at the same time that it sustains and counteracts the over setting force of the top weight, and the surplus power beyond these two will right the ship.

It is this surplus power, beyond the two effects counteracted, which gives stability to the ship; its measure is called "the measure of the stability of the ship," and the art of the ship builder is to make it always just as much as is wanted for this purpose, and no more, for more is of itself an evil, and defeats other good points.* From this examination of the sources of stability and instability in a ship, and of the part taken in that stability by the under-water body, by the shoulders, and by the weight carried, one is prepared to seek exact measures of the effect of each, and to obtain from them, as a result, the measure of the stability which is achieved. *[The surplus power of the shoulder measures the stability.]*

To measure the upsetting force of the under-water body, its volume is measured, after which its power is found by taking the weight of an equal quantity of water or its buoyancy. Then assume that power as applied at its centre of action, commonly called "centre of gravity." Draw a line directly upward through this point, and call that "the line of action of the upsetting force." Mark the place where this line cuts the water. It is on this water-line that the comparison between the forces causing stability can be most directly seen, and it may be called "the line of comparison." *[Measure of power.]*

To measure the righting force of the shoulder, in like manner measure its volume, then find its power by taking the weight of an equal quantity of water, or its buoyancy, reckon that as applied at its centre of action (centre of gravity), draw a line directly upward through this point, and call it the "line of action of the righting force." Mark the place where this line cuts the water, *[Force of the shoulder.]*

---

* Too much stability is almost as great an evil as too little.

28   POWERS OF SHOULDER AND UNDER-WATER BODY.

and here its action may be compared with that of the other two forces.

Top weight.   The third force has been already measured, it is the top weight placed on the vessel, and it is 12 tons for each foot of length. To compare this with the others, let fall through its centre of gravity, its line of action, which cuts the water at some place intermediate between the other two. Mark its place. The water-line now shows the three points of comparison desired.

Stability of form.   Of the three first, compare the upsetting and righting force of the body and shoulders. They are on opposite sides of the middle of the ship, they counteract each other. In the case under consideration, one, the upsetting force, is much larger in quantity than the other, the righting force,—larger in the proportion of 3 to 1; but the smaller force acts more advantageously than the greater, so much so, as to overpower it, because its centre of action is four times farther from the centre of the ship on one side, than the point of action of the other. The combined result, therefore, is in favor of the righting force, and the ship has stability, and will right itself. If the other force had preponderated, it would have had instability and have overset, even without a deck load. The question now remains: How much stability has it? In other words, how much top weight will the ship carry, and how high?

The surplus righting power measured.   To find this, multiply the volume of the under body by the distance of its line of action from the centre, subtract it from the righting force multiplied by the distance of its line of action from the centre; the balance, in figures, shows the balancing quantity of force the shoulder is able to carry. This may amount to a weight of 12 tons, multiplied by the distance of the line of action of the top weight from the centre of the line of comparison. If this be so, the vessel has stability enough, not to be overset; if, on the contrary, the surplus is less than this, the vessel will be overset. This surplus sustaining power, however, is the measure of stability. But, in this calculation, all consideration of the effect of the weight of the ship itself, either in oversetting or in righting, has hitherto been omitted. It may happen, and does happen in practice, that the weight of the ship alone, without a deck load, is enough to upset her. In such a case, the weight of the deck load must be treated as the whole weight of the ship, and the point of action of this weight must be taken in the centre of action of the sum of all the weights of all the parts of the ship and her equipment. In this view of the case, one must substitute, in the calculation, the total weight of the ship as well as the weights on it, instead of the deck load, and must examine the height at which the whole of these could be carried without upsetting.

Comparison of Ships of different magnitudes.   This height is taken as a convenient way of estimating the surplus righting power of the shoulders of the ship; because, in comparing different ships, one may, without reference to their weights or displacements, compare their righting powers or stability by the height above the water, at which they have power to carry their own weights. For example, a ship which has power to carry her own weight 6 feet above the water, and another which has power to carry her's 3 ft. out of the water, may be said to have relative stabilities of 2 to 1; but, if the magnitude of the ships also be considered, and one is double the bulk of the other, and has power to carry its weight twice as high, the absolute stability of the one may be four times that of the other, although their relative stabilities, reckoned by height alone, are as 2 to 1. The upsetting power of the bottom of a ship and the righting power of the shoulders are, therefore, the two rival forces which continually oppose one another. *Surplus righting power measured.*

These two forces depend entirely for their quantity, their proportion, and the manner of their action, upon the forethought, knowledge and skill of the *designer* of the ship.

The proper balance of these forces, in the design, makes the ship a good or bad carrier of top weight, and the height at which it can carry all its weights, is a point of the greatest value in every ship, and in men-of-war especially. If a considerable mistake be originally made, it is scarcely possible to correct it by anything short of re-building the ship.

CHAPTER IX.

ON THE PROPORTIONS WHICH MAKE A STABLE OR UNSTABLE SHIP.

Effects of the different parts of a Ship in giving stability. — In framing the design of a ship, few things are of greater importance to be clearly seen, and unceasingly kept in mind, than the effect of the bottom to diminish, and that of the shoulder to increase, power to carry top weight. In order to give a ship this good and indispensable power, it is important that the Naval Architect should not, for a moment, lose sight of the contrary nature and tendency of these two forces. It is from the omission of, or inadequate consideration given to these two effects, that crank, unstable and unseaworthy ships have so often been built.

Crankness. — Crankness was a general fault of ships built in the early part of this century. It means two things: inability to stand upright, and facility of being upset by top weight. The cause of crankness is often supposed to be shallow draft of water, which would be cured by deeper immersion. This is a radical error; there is no more common source of crank ships than this general impression. The contrary is the truth.

Crankness of old Ships. — Take a square ship, like a box, filled with light material, so as to sink no deeper than one-fourth part of its breadth, it will stand upright well; fill the same with heavier materials, so as to sink it to double that depth in the water, it will immediately turn bottom up. This is a very common proportion of draft to breadth, especially in old ships. It is sufficient to make a bad ship. But it is necessary to understand how to build crank ships and ships not crank. As a general rule, then, ships with a deep and large bottom and narrow shoulders, and ships with a straight, upright side, and flat bottom, and sharp bilges, will be crank.

Remedy for crankness. — In most cases, ships that are crank may be cured, by altering them so as to increase the breadth of their shoulder, without altering their bottom.* They may also be cured by lengthening them, so as to make them, with a given load, draw less water. Both plans have been tried with success.

Explanation of Table, p. 32. — The table at the end of this chapter, is given to show the limits of the power of a square built, wall-sided ship to stand upright under heavy and high loads. To each breadth there is a given height, up to which she can carry top weight, and the table shows with what proportion of depth in the water, to breadth, she can, or cannot, carry her weights above water; thus the table shows, that such a vessel, 36 feet broad and 18 feet deep in the water, cannot carry her weights if their common centre lie above the water, and that she would require to be 48 feet broad, to carry them just 20 inches above the water.

In this table, the figures 0 show that, if the whole weight carried were no higher than the surface of the water, the ship would, nevertheless, be incapable of standing upright, and would

---

* This is sometimes done by means of "sponsons."

PROPORTIONS OF STABLE AND UNSTABLE SHIPS.

either list over or upset. The figures show how high the centre of gravity of all the weights carried, including both the material of the vessel itself and the burden with which she is laden, might be raised above the water-line, without instability or danger of upset.

The value of this table is manifold, it shows how the extremely shallow, flat vessels on the Mississippi and other rivers, are able to stand up under their very heavy top loads, and carry enormous floating hotels, three and four stories high above the surface of the water. It is their small proportion of depth in the water, combined with their great breadth, which does it. It is this proportion which enables them to carry, not only their light cabins, but also their heavy engines, boilers, fuel and deck loads above the water. *Use of the Table.*

It shows the proportions for floating docks, which have to take ships of great weight, raise them high and dry above the water, and carry them steadily there. It also shows how high the centre of gravity of a ship may be, which a floating dock of given proportions can carry, taking into account, also, the weight of the floating dock itself. It shows how the shallow floating platforms of such contrivances as Clark's Hydraulic Docks, are able to sustain ships under repair, by using the right proportion of depth to breadth for a ship which has her centre of gravity at a certain height above the water.

This table enables one to see, also, how the square built, wall-sided, deep-bottomed ships, so often built by uninformed or careless shipwrights, turn out unstable and unseaworthy.

In using this table to judge of a ship or design, it must not be forgotten, that the case shown, is that of a box-formed or wall-sided vessel, nearly rectangular in shape; but it is nearly true, also, of a vessel slightly rounded off at the corners, and will be pretty exact for many large, capacious ships. It must be carefully borne in mind, that the table shows the extreme or upsetting heights to which the centre of weight must *not* be raised. The weights of a well trimmed ship, intended to carry sail well, should be kept so that the centre of gravity may be several feet under the limiting height.

It should be further noticed, that the length of the vessel is not given in the table. The breadth and depth being given, the length has *no* effect on the height at which the whole load can be carried. But length has everything to do with the quantity of weight which that ship will carry at the height in the table. Thus, a ship of 36 ft. beam, carries one ton for every foot deep; and for every foot in length, as many tons as there are feet of her depth in the water; therefore, it is to be remembered, that the weights carried at these heights are limited by the total displacement tonnage of the floating body. With these explanations, this table is a safe guide for the construction or judgment in regard to rectangular, box-shaped, or wall-sided, square bilged vessels.

POWER TO CARRY TOP WEIGHT

Heights out of water, up to which loads can be carried by vessels of square form and of different proportions of breadth of shoulder to depth of draft.

DRAFT.	BREADTHS.—Feet.													
Feet.	12	18	24	30	36	42	48	54	60	66	72	78	84	90
6	0	....	....	....	....	....	....	....	....	....	....	....	....	....
12	0	1.6	5.0	9.6	15.0	21.6	29.0	37.6	47.0	57.6	69.0	81.6	95.0	103.6
18	0	0	0	0.3	3.0	6.3	10.0	14.3	19.0	24.3	30.0	36.3	43.0	50.3
24	0	0	0	0	0	0	1.8	4.6	7.8	11.2	15.0	19.2	23.8	28.6
30	0	0	0	0	0	0	0	0	0.6	3.1.5	6.0	9 1.5	12.6	16.1.5
36	0	0	0	0	0	0	0	0	0	0	0	1.8.4	4.7.2	7.6
Limit'g heights of load in feet and inches.														0.9

Proportion of Breadth and Draft at which such Vessels will upset, if they carry Top-weight.

BREADTH.	DRAFT.	BREADTH.	DRAFT.	BREADTH.	DRAFT.
Feet.	Feet.	Feet.	Feet.	Feet.	Feet.
12	4.8990	54	22.0455	14.697	6
18	7.3485	60	24.4950	29.394	12
24	9.7980	66	26.9445	44.091	18
30	12.2475	72	29.3940	58.788	24
36	14.6970	78	31.8435	73.485	30
42	17.1465	84	34.2930	88.182	36
48	19.5960	90	36.7425	102.879	42

CHAPTER X.

THE METHOD OF MEASURING STABILITY.

1. The first method, is to determine how much top weight will careen the ship to a given large angle, say 14°—out of the perpendicular, or in war vessels 7°—in order to compare the stability of different ships with one another at this angle. Two methods of measuring Stability.

2. The second method, is to find the extremely small degree of careening which will be produced by an extremely small top weight.

By this investigation is discovered a curious quality belonging to crank ships, namely, that although a very small top weight may make them lean over a little, they may, nevertheless, offer great resistance to a great weight tending to incline them much. It is common to speak of such ships as being "tender" rather than crank.

The following are the successive steps:

1st. Measure the bulk of the under-water body, the ship being inclined on alternate sides to the given angle. First method.

2d. Measure the buoyant force of that bulk, taking 36 cubic feet of bulk for each ton of buoyancy.

3d. Find the place of the centre of effort at which this force acts, which is the point commonly called the centre of gravity of the under-water body. Next, through the point thus found, draw an upright line—cutting the water-line at some distance from its middle. Then measure this distance from the middle line of the ship.

4th. The line just measured, is called "the effectual distance of the upsetting force," and is multiplied by the number of tons already found as the measure of that force, the product is called "the momentum of the upsetting body." This momentum is taken as the measure of the upsetting force.

5th. Measure the bulk of the righting body, or shoulder under water when the ship is inclined at the given angle.

6th. Measure the buoyant force of the bulk of the shoulder, taking 36 cubic feet for each ton of buoyancy.

7th. Find the place of the centre of effort in the shoulder, at which this force acts, which is the point commonly called the centre of gravity, and is nearly two-thirds the breadth of the shoulder from the centre of the ship, or one-third from the outside.

Next, through the point thus found, draw an upright line, cutting the water-line at a point, of which measure the distance from the middle line of the ship.

8th. The line just measured, call the "effectual distance of the uprighting force," and multiply it by the number of tons already found as the measure of that force, this product call "the momentum of the uprighting force," and take it as a measure of the uprighting force.

THE METHOD OF MEASURING STABILITY.

First method of measuring Stability. 9th. Next subtract the smaller of these two momenta from the greater. If the upsetting force be the greater, the ship will overset in that position, unless some heavy weight be placed on the bottom, or some equivalent force be applied to prevent its oversetting, and such a force will, in order to be effectual, require to have a momentum at least equal to the difference.

10th. But if, on the contrary, the uprighting force be the greater, the ship in that position tends to upright itself, and can carry increased top weight, until this increased momentum becomes equal to the surplus righting momentum.

It is this surplus momentum either way, that is taken to measure the stability or instability of the ship.

11th. If the surplus righting momentum is taken and divided by the entire weight of the ship, the distance will be found to which this whole weight might be removed to one side, without upsetting. This distance is reckoned as another measure of stability. If, now, this last measure be taken and be divided by the line of the angle of inclination, the height will be obtained to which the whole weight of the ship might be raised without upsetting it; and this is a third measure of the stability of the ship, and is called the measure in height of stability of form.

It may be found, geometrically, by taking the former measure and erecting a perpendicular to the water-line, which will cut the upright middle of the ship at this height.

Thus measures of stability are obtained in three forms:

1st. Power to carry a given weight at a given distance out of the middle line.

2d. Power to resist a given heeling force.

3d. Power to carry the whole weight at a certain height above the water.

Second method. The second method of calculating the stability of a vessel, is to calculate all the quantities given above, for some extremely minute angle of deviation from the vertical position. This may be said to measure the resistance of the vessel to deviation from the vertical, whereas the former method measures her tendency to return to the vertical, after having been compelled to make a great deviation from it.

ELEMENTS OF STABILITY.

Elements of Stability. *Breadth.*—The stability of a vessel increases or diminishes enormously with its variation—whether the displacement remain constant, or, the draft remaining constant, the displacement vary with the breadth. In the latter case the height of the meta-centre varies as the square of the breadth.

Displacement.—If the breadth remain constant, the stability increases as the displacement decreases. And, since in that case the centre of displacement rises, the height above the water-line, to which the vessel's load may be carried, receives a further increment.

Draft.—Assuming both the breadth and displacement to remain constant, the increase or diminution of draft lowers or raises

THE METHOD OF MEASURING STABILITY. 35

the centre of displacement, and with it the meta-centre. It does not otherwise effect the instantaneous stability. Elements of Stability.

Stowage of Lading or Ballast.—The meta-centre indicates the height to which the centre of weight of the vessel may be brought without upsetting; and the amount of stability for very small inclinations is measured by the distance between the meta-centre and the centre of weight. The stowage, therefore, effects the instantaneous stability, in so far as it raises or lowers the centre of weight, and not farther or otherwise. As the meta-centre fixes an absolute maximum, which being reached, the vessel has no stability whatever, the weights must, in practice, be kept considerably below it, as the vessel must have reasonable stability.

Curve Bounding Plane of Flotation or Form of Water-line.—This element of variation may be considered apart from all others, and even independently of the proportion between length and breadth. Cœteris paribus, fine lines may reduce the stability, measured by the height of the meta-centre above the centre of displacement, to one-half what it is in the rectangular box.

Length and Lateral Stability.—This element may be always disregarded except in the mere calculation of actual weights, provided the same breadths and depths at proportionate lengths are maintained.

Weight.—This element is merely a factor, and is of no other account in the investigation of stability.

A vessel with nothing moveable in her, has her stability completely determined by the moments, on the water-line, of the three following forces, only two of which are independent:

1st. Her weight.

2d. The upward pressure due to her displacement.

3d. The force required to keep these in equilibrium.

Distinction between Heeling and being Listed.—A vessel is said to *heel* when she is pushed over by an extraneous force; on the removal of which she would alter her inclination.

She is said to be *listed* when she has found equilibrium in any position, other than upright; whether owing to an unsymmetric distribution of her weights, or to any peculiarity of form. A *list*, therefore, implies equilibrium (though unsymmetric;) *heeling* excludes equilibrium.

As applied to a vessel *heeling*, the meta-centre has no meaning, except to indicate how an alteration of the weights might be made to give equilibrium. As applied to a *listed* vessel, it has the same import as to a vessel floating upright. In both these cases, it affords practical means of comparing many different forms; especially where the variation to be considered is in the water-line.

CHAPTER XI.

POWERS AND PROPERTIES OF SHOULDERS.

Shoulders give Stability.

The sum and substance of what is known of the nature of stability is this: the shoulders *alone*, give to the ship righting or uprighting power. No other part of the ship can be so formed as to increase the righting power given by the shoulders. The righting power given by the shoulders is equally effective in squaring the ship to the water, whether it be still water or rough wave water.

Not so the under-water body.

The bottom of the ship, or the under-water body, can in no way help the ship to keep upright, there is no kind of bottom on which the ship can be said to rest in the water; the most that any under-body can do, either by shape or size, is to take less away from the stability given to the ship by the shoulders, than some other shape or size of under-body takes away. Size of bottom, therefore, or quantity of under-water body, lessens the stability of a ship, and has to be counteracted by the power of the shoulders. In short, bottom tends to upset the ship; so much so, indeed, that if it be large and powerful, it may take more than the whole power of the shoulders to keep it down, and prevent the ship from capsizing. A large under-body, therefore, weakens the effect of the shoulder, by the whole of its upsetting power.

Effectual Stability.

It is only, therefore, the surplus power of the shoulder remaining over and beyond what is employed to keep down the under-body, which is available for use in carrying a press of sail, or in supporting top weight out of the water. If there be any such surplus, it is a duty to find out how much there is, to see if it be enough to carry a press of sail, and enough also to carry top weight—then the ship may be able to do without ballast.

Ballast.

By ballast, in the general sense of the term, is meant weights carried under the water—in contra-distinction to weights carried above the water, or top weights. There are two ways of ballasting a ship; one is, by real lading, or stowing heavy weights under the water; the other is by putting weights, which are not parts of the lading, nor essential parts of the ship, low down in the ship for the mere purpose of helping the shoulders to carry top weight—(this latter is the old principle of ballasting.)

Weight placed under the water, in either way, may be said to have the following effects: first, by being under the water as far as the top weights are above it, it neutralizes the bad effect of these top weights, and balances them. In this way under-water weight assists the shoulders in carrying top weight.

There is another way of looking at the effect of under-water weight in giving stability; it aids the shoulders in keeping down the under-body. In this way, as well as in counter-balancing top weight, under water weight helps the shoulders.

Forces affecting Stability.

Thus it is that there are three agents in stability; two arising from the shape alone, and one from disposition of weights. The shape and size of shoulder give stability of form—the shape and size of under-water body give instability of form—what of the power of the shoulder remains beyond counteracting this under-body, is the true surplus stability, or measure of righting power, for that form. This surplus is all that can be used for navigating a ship and carrying her top weights. If more stability be wanted, it can be obtained by weight alone. All the weights of a ship, which have their common centre of gravity in the middle of the ship, just between the two shoulders, neither help the stability nor hinder it. Only weight placed below the middle of the shoulders gives help and increases stability; and if the centre of all the weights of the ship, cargo and ballast, taken together, fall above the water-line, the surplus power of the shoulders may enable her to carry sail—if not, there is no resource left but to lower the weights in her, or to place ballast in her bottom; in other words, to supply the defect of stability of form by adding stability of weight.

As, therefore, stability of form is that power which the Naval Architect alone can confer on his ship,—while stability of weight may afterwards be regulated by those who lade, and control, and navigate the vessel,—the form and action of the shoulders are the province in which the skill, contrivance, and forethought of the designer of the ship can be most powerfully and usefully employed.

<small>Stability of form rests with the Naval Architect.</small>

CHAPTER XII.

HOW TO GIVE A SHIP STABILITY WITHOUT GREAT BREADTH OF SHOULDER.

Breadth of Shoulder not always attainable. Breadth of shoulder, properly placed, gives power to stand upright, and to carry heavy weights above the water,—but cases often occur in which stability is sought, and breadth of shoulder denied. This may arise from local causes, such as the narrowness of a dock entrance, or from a wish to obtain certain other qualities which may be inconsistent with great breadth.

How to be supplemented. In this case, the Naval Architect has only the dimension of length given, and the question arises: How can length help him? How can it come in the place of breadth? To find out how this help may be given by length to defect of breadth, one must remember that if the two ends are made very fine, in pro- **Fine ends contribute but little to Stability.** portion to the middle body, they may be considered as having little effect in giving increased stability to the middle. But one may go farther than this, and state distinctly, that in all vessels with fine ends, very large portions of the two ends have merely stability enough to upright themselves, and have no power whatever to help the middle body to carry top weight. These two portions, therefore, may be taken as neutral parts of the vessel—neither helping the middle body, nor requiring help from it; and therefore, it simplifies the subject very much to leave them altogether out of the question.

Bow and stern considered in detail—See Plates. Suppose S to represent the form of the bow of a ship, and W to represent the stern of a ship, at the mean depth of water. The form S barely stands upright with its own weight. In like manner the form W barely stands upright with its own weight, and a very slight elevation and depression of weights would make them absolutely neutral.

It is plain, therefore, that these forms, if taken as types of a certain kind of bow and stern, do not effect the stability of the middle body either way. These two ends may be assumed as types of the "clipper" bow and stern. Suppose T to represent a "bell" bow—it will be found that instead of being any use, this "bell" bow is unable to carry itself, and would require help to a very great extent. U, on the other hand, taken as a type of the wave bow, has a powerful surplus stability at its deepest immersion, while only at its lightest immersion does it need help. V, which is the extreme of the "flare-out" bow, is unstable at all immersions, and worse than helpless. It is plain, therefore, that the Naval Architect need not trouble himself to seek much help from any of these bows—it is to the stern that he should look for any help he may want in supplying the needful stability.

The after-body may be made to give increased Stability. It is found, by long practical experience, that there exists a wide scope for obtaining power, stability and weatherliness to a ship of limited beam, by a wise design of the *after* body.

A crank, narrow ship, may be rendered stable and weatherly by a very moderate alteration to the bulk and form of the after body. The secret of success consists in uniting with a very fine line under the water, a very full line at the surface of the water.

HOW TO GIVE A SHIP STABILITY, &c.

The form Y has great stability at its two deeper immersions,—and is not deficient in stability even at its lightest. So great is its stability, that very few midship sections even compete with it in stability; and to most of them, it would, at its deepest draughts, impart an enormous increase of that quality.

There is another point in the art of Naval construction where the constructor can use the form of the stern with great effect— for the power of a middle body to carry top weight lessens as the draught of water increases. This is the same as saying, that in proportion as heavy top weight presses the vessel more down in the water, so does this very depth in the water diminish the power of the midship body to carry its top weight. The reason for this is already known to be, that bottom buoyancy increases both the quantity of the upsetting force and the advantage with which it acts. Now let it be observed how the after body can be used, so as exactly to counterbalance this defect of the middle body, and make good the stability of the ship in exact proportion to the increasing top weight which presses it down in the water. <small>Diminution of under-water part of after-body.</small>

The skillful Architect will carefully cut away bottom buoyancy from the stern of the ship,—this will enable him to make the run clean and fine, as he wants it to be. Nearer the middle, the stern may be of the form Y, and further aft, of the form X; and finally of the form W. Each of these forms has a growing surplus of stability over that necessary to support itself, in something like the following proportion to the increasing draught of the water: Y—3 when lightest, and 7 when deepest; X—3 at middle draught, and 6 when deepest;—and W only 4 when deepest, but negative at the two other draughts. The more, therefore, of a form approaching to Y, the constructor can put into the stern, the more powerful will be the resources he will have developed in the stern to aid the good qualities of the middle body, and to supply stability to do the work required, exactly at the time and in the manner where it is most wanted. <small>Examples of Stability so gained.</small>

Constructors of the old school will declaim against a full after body—they will insist on a fine run—but in truth, there is no reason why either should be sacrificed, in so far as concerns its practical use. On the bottom of the stern then, give the finest possible run—it is there where it is wanted, there alone where it is useful, so there give it to the utmost. Near the surface of the water, on the contrary, fineness of run is not only of no value to speed, but has many disadvantages of every kind. A wise constructor will seek there the stability he wants; there—the buoyancy may be taken in large quantity near the surface of the water; there—it may be obtained without impediment or increase of resistance; there—may be taken as much as is wanted to make the vessel a good and stable ship. A mine of good qualities is here to be found, hitherto comparatively unworked, and it has been unworked mainly on account of a vague, indefinite, but wide-spread prejudice, having no better basis than the old saying: "Cod's head and mackerel tail," (or make her as much like a fish as you can.) "Cod's head," meant simply the putting the fullness and bluffness required for stability, to carry sail, in the bow; and "mackerel tail," meant taking it away from the stern. <small>Full after-body above, and fine run below.</small> <small>Fallacy of "Cod's head and Mackerel tail."</small>

In those days, it was not known that putting fullness and bluffness in the bow, to create stability to carry sail, was putting it in a place to render that sail useless, for there it prevented that sail from carrying the vessel rapidly and easily through the water. Whereas, the application of Russell's wave principle, enables the modern Naval Architect to take away all that bluff buoyancy from the bow, where it does so much harm, by simply transferring as much or more buoyancy and stability into that part of the stern where, instead of doing any harm, it does good in every way—in every way, because it leaves the bow fine, of the form of least resistance, of the form of least disturbance, of the form of greatest speed; and it transfers to the stern heavy weights which would harm the bow, and it brings bulk where it gives room, buoyancy and stability.

Especially in Screw Steamships. Moreover, this room is given in that part of the ship where room is generally of the greatest value, both in a mercantile point of view and in ships propelled by the screw, in a mechanical point of view, for it is exactly a form of stern, extremely fine and clean below, which is best suited for the screw's effective action, while the buoyancy and room above are all required, in order to carry and counteract the great weights and mechanical forces due to the action of a propelling power in the after end of a ship.

Wave stern the best. A mine of unworked good qualities lies, therefore, in the after end of a ship constructed on the wave principle, if the Naval Architect will work it with diligence, and not be diverted from using its ample resources by the prejudice he may encounter. In the eye of one trained in the old school, the fullness of the new form of stern may be ugliness, but it is ugliness only to the uninformed mind, which cannot see in that fullness and capacity the virtues hidden in its ample bulk.

CHAPTER XIII.

HOW TO MAKE A SHIP DRY AND EASY.

There is, probably, no point in Naval construction subject to such variety of opinion as how to obtain ease and dryness in a ship, head to wind. There are several causes of this, and consequently counterpart causes which make a ship wet, uneasy and laborsome. <small>Ease and dryness.</small>

It is necessary to examine this subject, to arrive at just conclusions, because these same causes also make it either easy or difficult for a ship to ride at her anchors in heavy weather, or in a storm in the open sea, when lying-to. The qualities proposed for consideration are among those which it is most important to decide accurately, because they are those which enable a ship to survive in safety the perils of the sea. <small>Importance of the question.</small>

The first elements of riding easy, are form and size of bow above the water. Some thirty years ago, it was believed that a sea-worthy, comfortable, safe vessel, must have a high, wide, roomy, round, bluff bow, and that such a bow would enable a ship to throw aside every head wave, and rise high and dry above the sea. The idea was, therefore, that great over-water bulk and buoyancy was the grand consideration for securing the ease, safety and comfort of the ship. <small>Bluff bows.</small>

It must be admitted, that the example of the Dutch, and of many others, countenanced these opinions of the old school, and certainly any one who has seen how the Dutch fishing boats, and the pilot boats on the coast of Holland, ride out a storm on that dangerous and shallow coast, and ride safely over the breakers, would be apt to form a prejudice in favor of a buoyant, bluff bow. <small>Dutch vessels cited.</small>

There are many points to be admired in the structure of these craft, which peculiarly fit them for their special purpose; their bows are more bluff even than a circle, they recede inwards under the bow-sprit, so that they are the extreme and perfection of bluffness. But there could be no greater error than to take them as the type of sea-going ships, although it is a common blunder to fancy that the form which answers well for one purpose on a small craft, answers equally for all purposes on the scale of a large ship. This natural belief has, however, been the parent of the greatest errors in Naval Architecture—it is an idol of tradition. <small>Well suited to their special purpose, but no farther.</small>

The best constructors of the present day, hold a belief contrary to all this. It is believed to be the experience of all intelligent men, who have sailed in good vessels of the modern form, that the long, fine, hollow wave line bow, carried well above the water, rides easy and gently head to wind, when a full bluff bow could not live.* <small>Wave bow preferable.</small>

---

* A marked illustration of this fact occurred during a cyclone in the road of Funchal, Madeira, in March 1858. An English sailing barque, (built of iron,) with the long fine bow, as above, rode out the gale and heavy sea with the utmost ease and dryness, without striking any yards or spars, while the full, bluff bows went on shore and were wrecked, with the exception of the U. S. Frigate "Cumberland," which was only saved by the superior nature of her ground tackle and equipment, and a fortunate change in the wind at a critical moment.

Example of experiments.

Russell (the accomplished author and promulgator of the wave line theory) mentions a striking instance in which he illustrated this some twenty years ago. He says:—"I built four cutters of four large ships, all of the same dimensions, with four different shapes of bow, a wave bow, a straight bow, a parabolic bow, and a round bluff bow. I allowed the four captains to choose each his own boat in the order of seniority. The oldest captain took the bluffest bow, of course, as the best sea boat, and the wave bow was left for the last. In order to test their dryness and safety, head to sea, I had all four taken out together and forced through the water at the same speed by a steam tug. The speed was steadily increased, until at last the water was coming over the bows of the bluff cutter in such quantities that the trial had to cease in consequence of the head-sea pouring into her and filling her; the boat at the same time yawing about wildly, beyond the control of her rudder, and threatening to go down. All this time the crew of the fine wave bow, at the same speed, were dry, easy and comfortable; and so there was an end, in this case at least, of the prejudice that the full bow was the safe and dry boat."

Behavior of full bow in a seaway.

On a large scale, however, the circumstances may be very different. Nevertheless, observation of the effects of propelling vessels with full bows, head on to the sea, leads to the same conclusion; and it is thought that whoever has studied the behavior of a full bowed ship propelled by steam against a head sea, or riding at anchor, or laid-to head to wind in a storm, must have observed the following effects of the full bow:

First. It is true, that the fullness of the bow does cause the ship to rise over the waves, and does cause the bow to ascend well in the air on the coming sea. Unluckily, it rises too high and too far on the top of the coming sea, and it follows, that, when it reaches the top of the sea, a great quantity of this bluff bow is left high and unsupported in the air, and out of the water, so that, for a moment, one might see right under the fore foot and keel of the ship. The bow rises, but it rises too much—for in the next second, the unsupported body falls with a rapidly accelerating velocity, and descends headlong into the falling wave. Rushing down this slope, the bow by its momentum in falling, plunges deep into the hollow of the wave. It is there met by the rising face of the next wave, which again lifts it high in the air, and it again plunges heavily into the hollow of the next sea. It is this plunge into the succeeding sea which produces that violent shock that *no ship* can withstand for a long time. The English steamer "Great Britain," was an example of a vessel very fine below, with a great projection given to her above, under the idea of obtaining sea-going qualities; in her first trial, however, she received serious damage from a sea striking her in the manner above described. The same thing happens to a ship with a full bow out of the water when she rides out a gale; the bow receives from the ascending wave a rapidly ascending motion, till she comes to the top of the wave, and then, going over the crest, the whole weight of her unsupported, overhanging bow pitches down into the succeeding hollow; half buried, she is brought up with a violent shock in the following

HOW TO MAKE A SHIP DRY AND EASY.

sea; and so she goes on, scending and pitching violently, over every crested wave.

It will be seen that such a vessel *cannot* make much headway through the water, pitching and scending on every wave; the force propelling the vessel no longer goes to speed. It goes towards driving her up on the ascending wave, and down on the descending wave, and each heavy stroke of the water on the immersed bow is just so much force expended in stopping the ship, straining the timbers and wasting the propelling power. Effective speed loses, therefore, as much by such a form as ease and security.

But, it may be asked, "how should a vessel move, if not up and down over the sea?" To which it may be replied, "up and down certainly; but not violently—as gently as possible." The movement up should be gentle, the vessel ascending just so much, that the rising wave may not enter the ship, and descending on the other side just far enough to recover easily and without a shock at the bottom of the wave. In short, the motion of the vessel up and down should be a little less than that of the wave, and a little slower, instead of what is described above—much more than that of the wave, more rapid and more violent. This desirable equilibrium is accomplished by a certain well proportioning of the bulk of the over-water part of the bow to the under-water part of the bow. When the under-water part of the bow is very fine, the over-water part of the bow must be made fine likewise. When the under-water part of the bow is full, the out-of-the-water part will have to be proportionably full, and this proportion may be best given by so arranging it, that the bow of the ship on the ascending part of the wave, and on the descending part of the wave, shall have nearly equal bulks, alternately exposed below the water line and immersed above it. *A vessel should ride gently.* *Bow out of water being proportioned to bow under water.*

It is to be observed, however, that at the bottom of the wave the way of the ship exercises more force upon the approaching wave, to bury itself, than at the corresponding point of the top of the wave, to rise out of the water. It is right, therefore, that the out-of-the-water part of the bow should be fuller than the under-water part—just enough to prevent her taking in a sea over the bows. *But a little more full.*

It is thought that a bow like the steamer "Great Eastern," accomplishes this purpose perfectly in all weathers; her bow, however, would even in modern practice, be regarded by many as too fine above the water. *Example of the 'Great Eastern.'*

It is evident, therefore, that this approximate equality of fullness of bow above and below water, has a tendency to make the sides of the bow between wind and water nearly straight, and also nearly vertical. *Sides of bow should be nearly vertical.*

To this kind of bow, there exist two in marked contrast, termed the "flare out bow," and the "tumble home bow." The "flare out bow," is often called the "clipper bow;" and there is another kind of it, formerly called the "bell bow;" and midway between all these, is another sort of bow, neither tumbling home, nor tumbling out, this may be styled "the upright bow."

HOW TO MAKE A SHIP DRY AND EASY.

Bell bow.

The "bell bow" was a favorite form with the builders of the packets trading between New York and Liverpool, thirty years since, before the mail steam service of the Cunard and Collins lines ruined that trade. It was a fancy of the builders of those fine ships—to give the bows a form somewhat resembling a church bell inverted, the swell outwards, or "flare out," as it is called, beginning about the "light line," and flaring out all around, to the top of the bulwark, so that the forecastle occupied, as it were, the mouth of the bell.

There was, no doubt, something graceful and majestic about the aspect of these great bows, swelling out above the breaking waves, bearing up the bow of the ship by their buoyancy and then coming down upon the sea with such overwhelming force as to dash the waves in wild spray all around. It was a grand sight, but a costly one. The bow, no doubt, buffeted the waves triumphantly, but meanwhile the vessel was engaged in other work than its duty. Its business was to have gone, not up and down, but forward, and this the bell bow hindered, and expended useful force in unnecessary but magnificent struggles. A bow was wanted, that should elude the waves and pass them, escaping its enemy not fighting it.

Clipper bow.

The clipper bow was next introduced, to accomplish the design of the bell bow, without involving its defects. Believing still in the advantage of a large flaring-out bow, the inventors of the clipper bow endeavored to obtain the supposed advantages of great buoyancy, without the impediment produced by so much immersion in the water as the bell bow involved. "For this purpose," said they, "let us bell the bow laterally only, but draw it out longitudinally into a fine point; thus we shall preserve its bulk, but improve its shape."

Hence the fashion came in of prolonging the bulwarks of the ship at the level of the upper deck a great way forward, even 10, 20 or 30 feet in front of the actual ship, and there they were drawn out into a fine point above, and joined to the real ship about the water-line, everywhere with a kind of hollow flaring outside. This system certainly mitigated some of the evils of the bluff bell bow, and a large volume of buoyancy in the upper part of the bow, enormously in excess of the part of the bow in the water, was obtained. Two classes of vessels have been much distinguished for the extent of this clipper bow.

Messrs. McKay, of Boston, and Messrs. Hall, of Aberdeen, Scotland, introduced it largely into practice, and their vessels have had remarkable success in many respects. This success, however, may fairly be attributed not to this flare out bow, but to very different qualities of a practical and real kind.

Flaring bow not recommended.

But, imitators are apt to copy defects in the model they admire, rather than imitate its merits, and the imitators of the clipper bow have copied its worst points.

It must always be regarded as a bad quality to have on the sides of a ship large over-hanging projections, whether they "bell out" or "flare out." It is enough to say, shortly, that they injure speed, and that they give uneasy motion to a ship, and that overhanging surfaces generally strike the water violently and uneasily. There can hardly result any good in a large, flaring

HOW TO MAKE A SHIP DRY AND EASY.

out bow, whatever its shape may be; and it must be paid for in weight of material, in want of strength, in resistance to speed and in uneasy motion.

The most plausible recommendation of a "flare out" bow, is, that it throws the water off, and makes a ship dry. This is true, in a certain degree, and in certain circumstances, but they are not general, and rarely belong to the cases now under consideration. If a vessel of fine and upright form in every respect, have a slight "flare out" given to it at the top, it will turn over the tops of the waves, and prevent some spray from coming on board; but if it really strike solid water instead of spray, it will do so with such force as to send that water into the air in large quantities, and if the vessel really take in green water over the "flare out" bow, the danger to the ship produced by the mass of water in that place is so serious, that no imaginary beauty can even justify such a defect. *Flaring bow keeps off sea and spray, but may ship a green sea, and is then dangerous.*

It is believed that a much better form of bow is the nearly upright, or say the "tumble home" bow, provided the construction of the other part of the ship will admit of it. *The upright or "tumble home" bow make a drier vessel.*

A dry vessel is made, not by a "bluff, overhanging" bow to bruise, beat and buffet the waves, but by a long, thin snake-like bow, to elude the waves, to pass through them so as never to break the water at all; in short, a bow so formed as to offer the minimum resistance to the passage of the vessel through the water, not in one direction merely, but in every direction all round the bow, above and below.

According to this fashion, the full projection of the bow should be on the water-line; that, alone, should first penetrate the waves, and the way being thus prepared, the others should gently follow it. The top sides of the vessel, therefore, should "tumble home," and the whole be rounded off so beautifully and smoothly that nothing should either catch the water, stop the sea, or break it. Observation and experience show that such vessels are the dryest and the fastest in bad weather, as well as the easiest sea vessels, and above all, the safest. If green seas ever come over the bows of such vessels, they have a much smaller quantity to take in, they hold much less, and what little does come in, is much further back, and consequently, much less injurious.

The "tumble home" bow has never yet become "a fashion," but vessels have been built with it, which have proved themselves so much the better for it, that it is thought that ultimately it will be generally adopted. For speed, easy riding at anchor, ease in a gale of wind, or safety from shipping a sea, no other form is equal to it.

There are two exceptions, however, to this reasoning: A very small boat must have a large body above the water, if it be an open boat, to prevent its being swamped, or filled with water; and where buoyancy cannot be given by other means, it may be given by flaring out. *But "tumble home" bow is not suited for open boats, or heavy weights forward.*

Another case is, that where the bow of a vessel is filled, as, however, it ought *not* to be, with extremely heavy weights, corresponding buoyancy must be placed on top of the bow, to make it rise to the waves. This is true, but it cures one evil by means of another.

Capacity under water. No vessel designed for great speed, ought to have much capacity under water in the extreme bow, and in no case should that part of the vessel be occupied with heavy weights.

CHAPTER XIV.

ON LONGITUDINAL STABILITY.

If, in conformity with the maxims growing out of the considerations in the preceding chapter, the flaring bow (which causes a ship to pitch high, scend deep and make bad weather, by striking heavily on the sea) is removed, a long step will have been taken towards making her easy and dry, and many common causes of unseaworthiness are thus got rid of. But having got rid of some obnoxious features, there still remains some arbitrary matters, a choice of which goes far to enhance the good qualities of the ship, or to improve them. *Flare of bow not the only possible defect.*

The over-water bulk above the water-line is got rid of; but in the choice of the proportion and form of the water-line itself, much has to be settled on which may improve or deteriorate the ship. It by no means follows that a ship without overhanging or flare out bows is either an easy, dry, safe or steady ship, whether riding at anchor or going through a heavy head sea.

The form of the water-line itself, is a powerful agent in ease and sea-worthiness. *Form of the water-line is material.*

The power of the sea to lift a ship's bow, and the force with which, when lifted and left by the wave, that bow falls down into the hollow of the succeeding wave, depend on the form of the water-line and on the place in which the water-line allows the weights of the ship to be carried. With regard to the best formation of water-line for the pitching and scending of the ship, it may be inquired, first, how shall the water-line be formed so as to make the ship ride and drive easiest through the sea? understanding by "easy," that she shall rise and fall gently, slowly and not far. Fortunately, there is a measure which tells this exactly. The power of the slope of a rising sea to raise a ship is measured by the same elements and methods as those by which we measure the power of the water to support the ship sideways, against the depressing power of her canvass on the lee side, or enable her to carry a heavy load upon one side.

The power of a head sea to lift the bow of a ship is, therefore, measured in the same way as lateral stability; only the elements are reckoned lengthwise, instead of being taken across the ship. To proceed then to the consideration of measuring the tendency of the sea to raise a bow, which has been depressed under its natural water-line,—let it be imagined that the bow is pressed under-water by a displaced weight moved from O, the middle of the ship, and placed in the bows at W. This weight presses a wedge-like part of the bow into the water, and raises the stern out of the water. The wedge of immersion of the bow acts by its buoyancy at its centre of effort with a righting force proportioned to the bulk of the wedge, and to the distance of the centre of effort from the point O. It acts in length just as the wedge of lateral immersion does in width. Its raising power is to be found, therefore, by multiplying its volume into the distance of its centre of effort from O. The power of the sea to lift the bow of a ship depends, first, on the bulk of the bow immersed, and *Measurements of longitudinal stability.*

Scending. then upon the length of the bow, on the fullness of the water-line and on the place of that fullness. If a water-line be full forward, it will have great lifting power; if fine forward, small lifting power.

Pitching. Having measured the power of a sea to lift a given bow, it is necessary to measure the force with which a lifted bow, when left unsupported, falls down upon that sea. That depends upon the weight left unsupported, and on the point at which it acts. If the weights lie far forward on the bow, it will fall with great force into the water, descending with a speed proportioned to its distance from the centre, and plunging to a depth proportioned to the square of its falling velocity.

Effect of distribution of weight. From these two considerations, it is plain that a ship in scending, will be raised out of its horizontal seat on the water-line in a very high proportion inverse to its fineness of bow, and that in pitching, the ship will plunge less in proportion as the weights left unsupported, are removed from the extremities towards the middle body of the ship. The importance in trimming a ship, of so distributing her weights as to diminish their effect in causing her to pitch heavily, is therefore evident. A

Effect of the form of the bow. bow fine at the extremity, but taking fullness farther aft, makes a ship much easier than one which is leaner aft and fuller forward.

The form and proportion of the stern has not here been noticed, since the stern of a ship has not to be driven against a sea. In all ordinary practice, the stern of a ship is so sheltered from the sea, that overhanging form and unsupported weight, which would be a source of insecurity in the bow, may be tolerated with safety and convenience. What is said of the bow, may therefore apply only to the stern in a very modified form.

CHAPTER XV.

ON THE QUALITY OF WEATHERLINESS, AND HOW TO GIVE IT.

The nature, cause and cure of crankness or instability have been considered, and it is now known how to confer upon a ship the virtues counterpart to these vices—how to make her stable under every ordinary condition of load, and further, how to give her power of shoulder to stand up stoutly and carry a heavy press of canvas in a stiff breeze—a quality which is therefore called stiffness.

This quality of stiffness under sail, or uprightness, requires weatherliness—a virtue the opposite of leewardliness. A leewardly ship is liable to be driven with the wind, though her head may be laid in an opposite direction. By leewardliness, ships drive broadside on towards a lee shore, instead of lengthwise through the sea, and so lacking weatherliness, are lost. Weatherliness.
Danger of leewardliness.

Next, therefore, after stiffness, comes weatherliness to go in the direction intended—to make headway across the wind, and against the wind, instead of driving broadside to leeward.

This quality is to be obtained by considering an entirely different aspect of the vessel from that hitherto examined. The ship has been viewed through her breadth merely, she must now be looked at through her length and depth. Weatherliness, and how obtained.

The full length side view of a ship, as she sits upright in the water, presents a much larger extent of surface than the cross view of her breadth.

A ship of 36 feet beam, and 18 feet deep in the water, may have 648 square feet of immersed cross section; and, in order to force the vessel through the water, those 648 square feet of midship section must be pushed in the direction of the length of the vessel. This the propelling power of the ship must do; and it may require a pressure of 30 lbs. to push each foot through the water at the rate of ten (10) miles an hour. If, therefore, the sails of the ship had enough pressure of wind upon them to give this force of 30 lbs. for each square foot of section, forcing the vessel lengthwise, the vessel would go ten miles an hour before the wind.

This, however, is *not* the thing wanted; for the sails may be so trimmed, and the vessel's head so laid, that by means of the obliquity of the sails to the course, and of the course to the wind, the force of the wind shall partly force the vessel the way of the wind, and partly the way her head lies. It is the business of the Commander to lay her head in the proper direction, and of the officer of the deck to see that her sails are trimmed to the proper angle: but it is the Naval Architect's work to see that the form of the vessel prevents her driving to leeward. A ship may traverse the wind.

It must be understood, therefore, that when the ship does not run straight before the wind, but lies obliquely to it, the force of the wind acts in two directions. Partly it forces the ship its own way, or to leeward, and partly it forces her in the direction in which her head is laid, or to windward.

The ship therefore goes in two directions—partly to leeward and partly in the direction she is heading. The practical question is, how to make the first as little as possible, and the second as much as possible—or to make the ship weatherly. This the Naval Architect has to do.

The means by which weatherliness is given, consists in interposing the greatest possible obstacle between the leewardly part of the wind and its effect. The ship must be so constructed that it will be hard for her to drive to leeward, and easy for her to go to windward. How then, can she be made hard to drive to leeward?

Weatherliness depends on large longitudinal section. The antidote to leewardliness is large longitudinal section—648 feet of cross section are to be driven in the course of the ship. There must be much more than this, and as much more as possible, in the other direction, at right angles.

If the constructor can put six times 648 feet between the ship and her going to leeward, the ship is made six times as hard to drive to leeward as to drive to windward. By this means, it is contrived that the ship's progress to leeward shall be very small in comparison to her progress to windward, even when the sails are so trimmed that there is as much force pushing her the one way as the other.

Example. In considering weatherliness, therefore, he has only to see by what means as great a surface as possible can be interposed in the water, so as to prevent the ship being forced to leeward. If the ship can be made six times as long as she is broad, and preserve her depth below the water all the way to an average of 18 feet—or say 17 feet at the bow and 19 feet at the stern, which is the same thing—then the ship has a longitudinal section six times 36 feet, or 216 feet long, by 18 feet deep, presenting on the whole a resisting area of 3888 feet.

Thus it is, that a considerable excess of length beyond breadth is necessary to give weatherliness, and therefore it will be plain, that unless adequate length be given, all the stiffness to carry sail, for which so much breadth of shoulder has been given, will be thrown away—because, if the ship can carry sail merely, and that sail only force her to leeward, it is useless. Stiffness, therefore, or breadth of shoulder, must have length to back it, or it is worthless.

The area of longitudinal section to give weatherliness, must bear a due proportion to stiffness, and to area of cross section. Stiffness measures power to drive the ship—cross section measures the force necessary to drive it ahead, and longitudinal section measures resistance to being driven to leeward.

Another element which comes in to assist weatherliness, is the ease with which the fine shape of a vessel will permit her to be driven endwise through the water. It is a fact, that some forms of ships are so well contrived for this purpose, as, by sharpness alone, to reduce the power necessary to propel them,

ON THE QUALITY OF WEATHERLINESS. 51

to one-twelfth of what it would be, if they opposed to the water simply a flat bow.

It must now be seen what these qualities would be in a ship so designed. The area to resist leeward motion has already been made greater than that resisting forward motion, in the proportion of 6 to 1; and if the form be so fine as to reduce the resistance to forward motion still further in the proportion of 6 to 1, the combined effect will be in the proportion of 36 to 1. This would be a very successful achievement, for it would reduce the loss of motion, by leewardliness, to a very small quantity. *Fine form assists the longitudinal section.*

It is generally reckoned, that the extent of sail which a ship can carry in a fresh breeze, may be six times the area of her longitudinal section in the water. This, in the size of ship, taken, as an example above, would give an area of sail equal to 23,328 feet (or 3888×6.) Now this area of sail has got to propel the vessel with an effective force of 30 lbs. to each square foot of cross section; and as there are 36 square feet of canvas for every foot of driven section, the result is, that 30 lbs. divided over 36 feet, or 5-6 of a lb. is the required force of the wind on a square foot. Therefore it is plain, that a little less than one pound pressure on each square foot of sail, effective in the direction of the vessel's course, would be necessary to propel her ten miles an hour, and this a very moderate force of wind would accomplish. *Example of the effect of sail in headway.*

But an equal force with this would, with a given trim of sail, be pressing the ship to leeward. The effect of this other force would, however, be expended on six times the area, and that area has six times as much resistance to leeward as ahead. Under these circumstances, the motion through the water being as the square root of the force, the leeward motion would be to the onward motion as the square root of 36 to 1, which is of course, 6 to 1. *Effect of sail in leeway.*

The result is that the ship is driven six miles forward while she is driven one mile to leeward; and such a vessel would be an ordinary full, but *not fast nor weatherly* ship. *Resultant motion.*

There are three ways in which the Naval Architect can improve the weatherliness of this ship. He may diminish the area of the cross section—he may fine the shape of the ship so as to offer less resistance—he may increase the area of longitudinal section, and give increased resistance to leeway, by increase of length, or increase of depth—or he may do any or all of these things at once. *How the Naval Architect can make a ship weatherly.*

The process stated above, assumes that the Naval Architect is at liberty to give sufficient longitudinal area by the disposition of the body of the ship—that is, that he can have such a draft of water, and such a length of body as he may select. When his ship is not of suitable dimensions, he has to resort to various expedients. If he has not depth of water enough naturally in the body of his ship, he has to add timber, or deadwood, as it is called, for no other purpose than to increase the weatherly section of the ship. When he adds this on the bottom it becomes keel, or false keel—and is often carried to a *Special expedients, as deadwood and false keel.*

great extent. If he cannot get enough in this way, he adds further deadwood, in the shape of stern and cut water; and to assist and balance these, he adds as much deadwood as he can in the run before the rudder. It is thus that vessels with a small body may obtain a great weatherly section—and racing vessels, yachts and clippers are frequently built in this manner, to so extreme an extent as to be nearly all deadwood and keel, and little or no body. A vessel of this sort becomes a mere racing phenomenon. But, nevertheless, by extending deadwood in every direction, before, abaft, and below, extraordinary weatherliness may be obtained at the sacrifice of capacity.

Lee-boards. When these arrangements fail, or cannot be applied, there remain many expedients for securing weatherliness. The lee-boards of the Dutch craft attain this. On the shallow sandy coast of Holland, no deep keel is possible, and therefore, the Dutch vessel at sea would drift to leeward for want of depth of body; to provide against which, she carries on her lee side a large flat board of enormous area, which is let down into the water in such a manner that the whole of the board must be driven on the flat side, leewardly through the water before the vessel can make leeway. One of these lee-boards is carried on each side of the vessel, so that either side when it comes to leeward, has its own lee-board for alternate use. This is the Dutchman's substitute for windwardly section, of which his small draft deprives him.

Sliding-keel or centre-board. Another substitute has been used, but it is not thought with any very great degree of success as yet. It is termed a "sliding-keel," or "centre-board," and is formed by providing a hollow, upright aperture in the middle of the vessel, in which a large flat board is contained, so that it can be lowered through a slit in the bottom, into the water, and have to be driven broadside to leeward.*

These, however, are expedients merely in the last resort, when the Naval Architect is denied the means of giving his vessel a due proportion. If due length can be given, it is much wiser to obtain weatherliness by proper length and fine form, than to seek artificial expedients, either in lee-boards, or centre-boards, or in exaggerated deadwood; but of none of these expedients should the Naval Architect be ignorant; and it is better to obtain weatherliness by all, or any of them, than to have a leewardly vessel.

Another example. In these days, a sailing ship of ordinary form, is generally about six times as long as broad. To drive her at the rate of ten knots an hour through the water, requires about 48 lbs. of force for each foot of her midship or greatest cross section. Suppose the vessel has 100 square feet of immersed cross section, requiring a force of 4800 lbs. to give her headway, and that the sails are placed at such an angle, that they press equally forward, and over, or so that there shall be equal forces causing headway and leeway, there will then be a force of 4800 lbs. causing leeway; and this force is spread over 600 square feet, forming the immersed longitudinal section of the ship. On each square foot of this section, there will be, therefore, only one six-hundreth part of 4800 lbs., or 8 lbs. per square

---

* Centre-boards are only used in small vessels.

ON THE QUALITY OF WEATHERLINESS. 53

foot of section. This 8 lbs. will cause a leeway of less than 2 knots. This example shows the great advantage obtained in sailing vessels by large hold of the water. It is plain that, by giving this vessel greater length and depth, her resistance to leeway might be doubled, so that the force causing leeway would be divided over double the area, and be reduced to 4 lbs. per foot, instead of 8, and this 4 lbs. per foot would only give a leeway of 1.25 knots per hour.

Table explained. The following table shows what happens when the sails are set at an angle of 45° to the course of the ship, and the wind is right abeam. In these cases there is equal pressure along the ship's course, and to leeward. The lesser leeway arises from two causes: the greater area of longitudinal section than of midship section, and the fineness of the shapes. It will be seen, that in the full form of vessel, the leeway, under the pressure that produces 12 miles an hour, is 2 miles an hour. Under a pressure of 1 lb. on the sails, the headway is 10 miles an hour, and the leeway 1.2-3 miles. This would be the case in a fresh breeze carrying all sail. When the vessel is proportioned for greater speed, with a greater proportion of length to the same area of resistance, and a finer form, the leeway is reduced and the speed increased.

Limitation of results. These calculations are made upon the supposition that a vessel's resistance to leeway is the same as that of a thin plate equal to her longitudinal section. But vessels with round bilges let the water pass underneath them from one side to the other more easily than a vertical plate, and so do all ships when they careen much. A little more leeway must be allowed for in these cases.

TABLE OF LEE-WAY AND HEADWAY.

The sails set at 45°, the wind a-beam. The force of the wind reduced to allow for obliquity.

Full form of Ship, with the cross-section to the Longitudinal section as 1:6. A square fathom of sail area (36 ft.) to a square foot of cross-section.

Finer form of Ship with cross-section to Longitudinal section as 1:8. 36 sq. ft. of sail area to 1 sq. ft. of cross-section.

Effective force of the wind per square foot of Sail.		Drifting force to leeward pr. ft. per foot of longitudinal sect.	Driving force ahead per foot of cross-sect.	Drift to leeward.	Speed ahead.	Drifting force to leeward pr. ft. per foot of longitudinal sect.	Driving force ahead per foot of cross-sect.	Drift to leeward.	Speed ahead.
Leeward. Pounds.	Ahead. Pounds.	Pounds.	Pounds.	Miles.	Miles.	Pounds.	Pounds.	Miles.	Miles.
0.2	0.2	1.2	7.2	0.8	4.5	0.9	9.6	0.6	6.0
0.4	0.4	2.4	14.4	1.1	6.3	1.8	19.2	0.9	8.5
0.6	0.6	3.6	24.6	1.3	8.3	2.7	28.8	1.1	10.4
0.8	0.8	4.8	28.8	1.5	9.0	3.6	38.4	1.3	12.0
1.0	1.0	6.0	36.0	1.7	10.0	4.5	48.0	1.5	13.6
1.2	1.2	7.2	43.2	1.9	11.0	5.4	57.6	1.6	14.6
1.4	1.4	8.4	50.4	2.0	12.0	6.3	67.2	1.7	15.8

CHAPTER XVI.

HOW TO MAKE A SHIP HANDY AND EASY TO STEER.

The first part of handiness consists of balance of sail; the second, of balance of ship; the third, of proportion of rudder. <small>Handiness.</small>

Unless sail be balanced, the ship will go just where she pleases, or just where the wind pleases, instead of towards her destination. If there be a sail on the forward part of the vessel only, the wind will force the forward part of the vessel to leeward, and she will drive head foremost. If there be a sail on the after part of the vessel only, her stern will go to leeward and she will drive stern foremost and head to the wind. In order that neither of these things shall happen, the sail on the forepart of the vessel must be so placed and proportioned to the quantity and place of the sail on the afterpart that they exactly balance one another in effect, so that neither one nor the other can prevail. This is virtually to take away from the wind all power of determining the direction of the ship, and a smart officer, by properly regulating this balance of sail, can keep the ship's head in any direction he pleases. <small>Balance of sail.</small>

This is balance of sail, but it depends on another element, namely: the balance of ship. The effect of sail at the bow may be exactly balanced by that at the stern; yet, nevertheless, there will be no enduring balance if the bow be more easily forced to leeward than the stern, for then the head of the ship would go around to leeward. A balance of sail forward and aft, and a balance of ship lengthwise in the water—the one called "trim of sail," the other "trim of ship"—the forethought of the Naval Architect must provide. To maintain this equilibrium, depends upon the ability and thoughtfulness of the Commander of the ship. <small>Balance of ship.</small>

It is thus only, that a handy ship is obtained and kept so. The sails *must* balance, the body must balance, and both must be kept together in perfect trim; while the seamanship of the Commander, the hand of the helmsman and the blade of the rudder do the rest. <small>Must be adapted to each other.</small>

Something more, however, can still be done by the Naval Architect to give the sailor complete command over his ship. The balance he has established is enough to deprive the wind of the control of the vessel and give it to the Captain; but, the ship may still require from him the exertion of very great controlling force when he wishes, in the course of his manoeuvres, to change the ship's head rapidly from one course to another. Balance of sail and of body will help him to do this, but it will not help him to do it quickly. <small>Something more required for quickness;</small>

To make a vessel very handy, and turn very quickly, her longitudinal section should be deep, rather than long, and when its extreme length is decided, its effective length should be diminished as much as possible by removing longitudinal <small>Namely, that lateral resistance should be chiefly in the middle and *not* at the ends.</small>

area from the ends, and placing it near the middle. Above all, much cut-water and fore-foot make a vessel unhandy and slow to come around.

It is better, therefore, to have deadwood aft than forward, but removed from both ends as much as possible. Rounding off the fore-foot and shortening the heel, are the most effectual ways to make a ship handy without injuring her other qualities; the effect of heel and fore-foot being to cause gripe, or resistance to turning, which is the contrary of handiness.

Effect of rudder. With balanced ship, and balanced sail, good trim and little gripe, not much can be wanting to handiness. The rudder must do the rest. The rudder is nothing but a power to control, it merely acts as a drag on one side, it always diminishes speed in turning the ship, and the cleverest helmsman is he who uses it least, the best ship is that which wants it least, and the best sailor is he who does most without it. A steersman always *The rudder should be powerful—but sparingly used.* yawing a ship about, steers badly. A ship requiring much helm is badly trimmed. Sails requiring much rudder, are badly set or balanced. Nevertheless, it is above all things necessary, that the rudder should have ample power—great power, seldom used. A ship that will run along for hours with scarcely a touch of the helm, is a ship well trimmed and sailed; but when needed, the rudder must be able to do anything, and to be able to turn a ship, short and sharp around, may save her in an emergency.

The rudder should bear a certain proportion to the ship's length. The way to give power to the rudder, is to proportion it to the length of the ship. A long ship requires a broad rudder. It is thought, that for every 100 feet in the length of a ship, she should have 2 feet of breadth with 1 ft. added. Thus a ship 100 feet long needs 3 feet breadth of rudder; 200 ft. long, 5 ft. breadth; 400 feet long, 9 feet, and so on.

Shape of the rudder not very material. As to the shape of the rudder, there is not much in it. Some say the top of the rudder is the most valuable part, for the water there has the most effect, and that the rudder should be widest there; others say it should be widest at the bottom, for that there width is most effective. Both are crotchets; but still there may be something peculiar in the case of some ships, to render both exceptionally true.

The fault of having the widest part of the rudder near the load water-line is, that there a rough sea may strike the rudder most heavily, there is less harm in making the rudder widest near the keel, for being well buried under water, the wave surface of the sea has less action upon it, and in bad weather the helmsman is not liable, as in the first case, to have the helm taken out of his hand. *But perhaps is best rounded off gently at both head and heel.* But on the other hand, the heel of a wooden screw ship may be, and probably is, her weakest part, and to put more strain than necessary on a weak place, is unwise to say the least. The best way is to have the widest part of the rudder near its centre, rounding it off towards the top and heel, the one to keep it from the force of the waves, the other to protect it from the ground in a narrow and shallow channel.

It will always be a delicate question about the quantity of rudder to be given to a vessel destined for any special purpose.

HOW TO MAKE A SHIP HANDY.

If the ship is always to be committed to wise hands, who will never use more than is necessary, it is safe to give plenty of rudder, leaving it to their discretion to use it as they may desire. Strictly speaking, a smart Commander should have just as much rudder as he can command and handle, because with powerful rudders, manœuvres can be performed which are impossible with small ones. To be able to turn very fast, will often give a ship the advantage of another; and in a contest for victory, or of sport for a challenge cup, ability to execute difficult manœuvres rapidly, is often in itself a source of success. A ship well in hand, is often better than one which is faster, but runs wild. Therefore, put into wise hands a powerful rudder. *Caution as to giving too much rudder power except to a discreet Commander.*

Unluckily, the Naval Constructor has sometimes the misfortune to have his finest work entrusted to the hands of a fool. In such case it would be vastly better, if possible, that the rudder should be proportioned to the capacities of the man, rather than to those of the ship, and the rudder should not be so powerful as to enable the ship to perform the intended manœuvre too rapidly, lest it should outrun the wits of the fool in command. The ship should be able to come around only just so quickly that the governing intelligence may keep pace with her movement, and see what she is going to do; only, therefore, to a quick witted man intrust a quick working ship.

For a slow ship and a slow Commander, a narrow rudder is best. The motions of the ship will thus be slow, they will give time for thought, and will be less likely to run the ship into mischief. For such ships, two-thirds the dimensions given above will be sufficient.

CHAPTER XVII.

OF BALANCE OF BODY AND BALANCE OF SAIL.

Handiness depends on balance of sail. It has been seen that handiness, or the ready obedience of a ship to the will of her Commander, arises out of the due combination of balance of sail, balance of body, and power of rudder; and it has also been seen how, without these, a vessel steers wildly and can hardly be controlled in her movements.

Ardency and leewardliness. When, either through want of balance of sail, or balance of body in the water, the ship shows a tendency to fall off or fly to, she is said to have two opposite defects, the first called leewardliness, the other called ardency. These defects must be corrected either by trim of sail or trim of ship. If not corrected, they must be counteracted by the action of the rudder; but, as the rudder is a sort of stop-water applied on one side, and in no case a help, speed is lost in the degree in which the rudder is used. The tendency to fly into the wind, and the tendency to fall off from the wind, or ardency and its opposite, require remedies of opposite kind. Ardency implies that the ship must always carry weather helm; want of ardency, that she must always carry lee or slack helm. Of the two evils, ardency is considered the less, and it is usual, therefore, to trim a ship so that she shall always carry *a very little* weather helm.

Centre of effort of sail should correspond with centre of lateral resistance. The point in the length of a ship, on both sides of which the sails balance, is called "the centre of effort of the sails," the point in the water on both sides of which, fore and aft, the body balances—or to put it otherwise, the point in the air on both sides of which, fore and aft, the pressure of the wind on the sails balances, and the point in the water on both sides of which, fore and aft, the resistance of the water to the leewardliness of the ship balances, these are respectively called "the centre of effort of the sails," and "the centre of lateral resistance of the ship."

In a state of perfect trim of sail and trim of ship, that is, when a ship is so perfectly balanced as to be neither ardent nor leewardly, requiring neither lee nor weather helm, the centre of effort and the centre of resistance would meet exactly in the same point of the length of the ship, and so the effort of sail and resistance of water, fore and aft, exactly counterpoise one another.

Unfortunately, there are but few ships so constructed as that this coincidence shall take place and be maintained at all speeds, and in all states of wind and sea and weather.

In a perfectly mathematically formed "wave line" vessel, the coincidence of the two has been found to be exact and perfect; but by a very slight deviation from the wave form, the perfection of this balance is at once deranged. In order to correct this want of adjustment, the centre of effort of the sails has to be moved forward; and in certain cases very considerably so.— (The English line of battle ship, "Duke of Wellington," for example, in which the whole of the masts and sails had to be placed forward from the true centre of resistance the space of 14 feet.)

BALANCE OF BODY AND BALANCE OF SAIL. 59

In every vessel built on the old system, this derangement of balance had to be taken into account, and allowed for, as an element in the original construction of the ship. Unluckily, it was allowed for by guess, merely; and, therefore, nothing was so common as to hear that a new vessel had to undergo an entire change of arrangements from the impossibility of managing her, owing to the centres of effort and lateral resistance not coinciding. In such a case the usual remedy (if the error was slight) was to rake the masts either a little forward or a little aft, in order to correct the balance of sail, or else to put on a little deadwood forward or abaft, or to add a tapering false keel, all for the purpose of restoring the lost balance; and when these expedients failed, the masts, or some of them, had to be shifted—an arrangement not only expensive but deranging to a great extent the interior economy of a ship of war. *Remedies applied.*

One of the great advantages of Russell's wave line system, is that the centres of effort of sail and of resistance of body coincide. It is impossible to adjust these two centres to a more perfect balance for practical use, so as to have a ship easy to steer, requiring little helm, quite under command, and handy, than by merely taking care that they coincide in the same point of length.

But this perfection of balance and handiness, is not to be obtained without an equal perfection of wave form. Every deviation from exact truth in the form of a ship, will exhibit derangement of balance. In exact proportion as any part of the bow is filled up beyond the pure wave line, the balance of sail and of resistance will be disturbed, and it will be necessary either to correct the shape of the body, or to remove the centre of effort of sail forward, or to shift the centre of lateral resistance aft. The reason of this seems clear. The wave form is the form of least resistance; the truth of which is practically shewn when the vessel (in deep water) shows no bow-wave, or breaks no water at the bow. Any untruth in the wave form, at once shows itself in broken water, or the well known wave at the bow, which is bad. The appearance of this obstacle to progress shows exactly where there is an expenditure of undue force, and it is this undue force and unnecessary resistance which deranges the centre of lateral resistance of the ship, and shifts it forward. Its tendency to do so increases with the velocity of the ship. It is to meet this shift of pressure and to counteract it, that the centre of effort of the sails must follow it forward. But no one can tell precisely, beforehand, how much any deviation from truth in the form of a ship will remove the centre of resistance, and, therefore, it is impossible to say what change in the centre of effort may be required to correct it. Unluckily, also, the deviation arising from incorrect form, varies with the speed, so that the difference which will restore the balance, is not the same for all speeds. No certainty, therefore, is to be obtained on this point, except by the preservation of the absolute truth of the wave form. *A departure from the wave form moves the centre of effort of sail forward.* *Deviation varies with the speed.*

There is another curious cause of deviation between the centres of effort and resistance. If a ship has a long straight middle body, she will have a tendency to ardency, arising from length alone. Even if the two ends be perfect wave-ends, a long *A straight middle body gives ardency.*

straight middle body will have this tendency to disturb their balance. Of this singular phenomenon of deviation, arising from length of middle body, an exact measure can scarcely be given; but an explanation will be attempted. A long, parallel bodied ship certainly loses her power to carry after-sail; and the explanation is believed to be, that a long ship, by the mere progress of its sides through the water, drags with it, and puts into motion by adhesion merely, so great a quantity of the water in its neighborhood, that at the last, when near the stern, the water has ceased to offer any lateral resistance, because it has already received the same motion as the ship itself. At the stern, therefore, there is little left to resist the ship; and so from lack of stern resistance, the after part loses power to carry after sail, and the ship becomes ardent. Even wave-ends, therefore, will not compensate for this fault of long middle body. If the inquiry be made as to what limit this extends, it may be replied, that in a vessel with 60 feet beam, and 90 feet middle body, it has not been sensible, but with similar ends, and 100 feet middle body, it has become a very sensible quantity. This deviation from the true form, while it is attended with mercantile advantages of capacity not to be regarded lightly, must be taken with its disadvantages.

Explanation. In order, thoroughly, to apprehend the nature of this fact, suppose a thin, flat board, moving edgewise through the water, and also pressed sideways by a force like the wind at right angles to it. Next consider what happens to a particle of water placed to leeward of this thin board. When the board first touches it, it has no leeward motion, but it immediately acquires it; small at first, but gradually growing, as the following parts of the board successively press it, and, as each succeeding part of the board finds the water already put in leeward motion, it follows that the latter parts of the board are in contact with particles already moving so fast to leeward, that, unless they accelerate their leeward speed, they will experience no lateral pressure from the water. Hence, two effects must follow; the after parts of the board will have less pressure on them than the fore parts, and also the after parts will be moving to leeward faster than the fore parts. This is the explanation of the ardency produced by a long, parallel, middle body.

Other causes of ardency. Another source of derangement between the centres of effort and resistance, will be found in any deviation from the water-line which may be produced in the change of shape in the vessel, as she heels over from the pressure of side wind. If a full part of the ship comes into the water on heeling over, that part will cause its own special resistance, and, so far as it deviates from the true form, will cause an excess of pressure at that point, and a derangement of the centres of balance: it will in fact make the vessel behave as if she had a curved keel, concave to the wind. Another reason why heeling produces ardency, is that it forces the centre of effort of sail to leeward, so as to make the masts exert a horizontal leverage, to bring the ship's head into the wind. Apart from heeling, there is also a small element of ardency in the case of fore and aft sails, from their centre of effort being invariably to leeward. Theoretically, the bellying

BALANCE OF BODY AND BALANCE OF SAIL.

of square sails should tend in the same direction, but it is believed that this cause is not appreciable in practice.

There is another cause of deviation, which must take effect in all vessels, of whatever form, but its action is slight, and is not, except in the case of very great length of middle body, of sufficient consequence to rank as an element in the adjustment of the centres—it is the resistance of the adhesive film of water on the skin of every ship. This adhesive film is scarcely a visible thickness at the bow; it increases uniformly with the distance from the bow towards the stern, where it is greatest; the invisible film seems to grow as it goes, by attaching to itself another and another outside film on each foot of progress, and, all added together, the entire film has a thickness of a foot or more on each side at the stern. This may diminish the lateral resistance, but it seems just enough to give the vessel that degree of ardency which is preferable to the smallest degree of the opposite quality; and when no other source of derangement than this remains, the Naval Architect may congratulate himself on having completed that part of his business satisfactorily. One other source of disturbance will always remain, which he can neither forsee nor prevent—the winds and the waves will always act irregularly on the ship. But when he has done the preceding parts of his work well, he will leave the mind of the helmsman, and the action of the rudder perfectly disengaged from all unnecessary work, and free to be disposed of in that cautious, ready and prompt counteraction of the winds and the waves, which is the business of the thoughtful and the watchful seaman. *Ardency due to adhesion of water.*

It is the duty of the ship-builder to make an exact calculation on the body of his ship, so that, when loaded in the water to its proper line, the lateral resistance to leeway shall be found at its proper point in the length of the ship. This central balancing point he calls "the centre of lateral resistance," and it should be a little abaft the centre of the ship. *Ship-builder must find the centre of lateral resistance of ship—*

His next duty is to make an exact calculation of the sails of his ship, so that the pressure of the wind upon all the sails may have its central balancing point rightly placed in reference to the centre of resistance of the ship. This is called "the centre of effort of sail," and should be so adjusted, as neither to be too near the bow nor too far from it. In some wave line vessels, it is necessary to place the centre of lateral resistance of ship, and the centre of effort of sail, precisely the one over the other, but in the forms of ordinary sailing vessels, it is found necessary to have the centre of resistance a little abaft the middle of the ship, and the centre of effort of sail a little forward of the middle. To bring the centre of resistance aft, a ship is generally trimmed two feet by the stern; and to carry the centre of effort forward, additional sail, beyond the quantity proper for a vessel on an even keel, is carried on the bowsprit. By these means, a distance of about one-twentieth part of the length of the ship may be placed between these two centres, and in most ships this is sufficient for the purpose. On page 63, is given a table shewing the distance to which trimming the ship by the stern will shift the centre of resistance backwards, and also the additional area of sail on the *And then the centre of effort of sail. Trim.*

bowsprit, which will suffice to place an interval of one-twentieth of the ship's length between the centres.

The trim depends more on the Commander than on the constructor. But, however exactly the Naval Architect may have designed his ship, the trimming of the sails and of the ship must remain essentially a part of the seaman's duty, because the trim of the ship is chiefly a question of stowage of cargo or disposition of weights, and as this is always varying in a steam-ship, and can always be ill or well done in a sailing vessel, the original constructor of a ship, however wise, can never dispense with the watchfulness and judgment of the Commander of the vessel.*

Centre of resistance may vary at different speeds. There is another cause for constant watchfulness as to trim, in the fact that most ships shift their centre of lateral resistance forward as their speed increases, and therefore require after sail to be diminished as the wind rises. The first duty, therefore, of an officer in command of a new ship, or one starting on a fresh trim, is to determine the proper balance of ship and sail.

He should shift weights forward and aft, until he finds the trim which will enable him to carry the proper sails, and having done this, should carefully study how the quantity of sail must be adjusted in the various degrees of strength of wind, so as to measure this balance. It is in this way, that a skillful Captain will often make a ship fast by trim alone, whereas an ignorant one will fail to find out the good points in a ship, because he does not systematically look for them, by studying her performance under every variety of trim at his command. In this way the Captain, even more than the constructor, makes the character of his ship.

Summary. The sum of what is known in regard to balance of body, and balance of sail and trim, is as follows:

The middle of the length of a ship is the balance point or centre of lateral resistance of a ship, if she be nearly at rest, drifting to leeward, and if she be on an even keel, with upright stem and stern post.

Trim and rake compared. Trim of ship by the stern, shifts the centre of lateral resistance from the middle towards the stern. An inch of trim to a foot of draft shifts the centre of lateral resistance abaft the middle by one hundred and forty-fourth part of the length of the ship. Or the excess aft, represented by a fraction of the draft amidships, (say one-sixth,) multiplied by one-twelfth of the ship's length, gives the shift abaft caused by trim. Raking the stern post and rounding the stem also shifts the centre of lateral resistance forward or aft.

Raking the stern post, shifts this centre forward one-quarter of the rake. Rounding the stem, so as to make it a quarter of a circle, shifts the centre aft by about one-tenth of the draft at the stem.

In the most ordinary shape of ships, these last counteract each other, and if the draft fore and aft be nearly equal, the centre of lateral resistance at rest is in the middle of the length.

---

* And the Commander should thoroughly understand the design of the constructor.

BALANCE OF BODY AND BALANCE OF SAIL.

To find what amount of trim will balance a given amount of rake of stern post, the following *approximate* formula may be used:

$$\frac{Trim.}{Rake.} = \frac{Draft\ Amidships.}{3\ times\ Length.}$$

Or else refer to following table, calculated from the formula:

Table showing rake of stern post required to balance a given trim or difference of draft forward and aft.

Rake of Stern-Post.			Inches of difference of trim to every foot of draft amidships.	Rake of Stern-Post.		
Inches of trim to every foot of length of vessel.	In fraction of length.	Decimal fraction of length.		In inches to every foot of the length of vessel.	In fraction of length.	Decimal fraction of length.
.17	1-72	0.01389	3.5	1.17	7-72	0.09722
.33	1-36	0.02778	4.	1.33	1-9	0.11111
.50	1-24	0.04167	4.5	1.50	1-8	0.12500
.67	1-18	0.05556	5.	1.67	5-36	0.13889
.83	5-72	0.06944	5.5	1.83	11-72	0.15278
1.	1-12	0.08333	6.	2.	1-6	0.16667

Inches of difference of trim to every foot of draft amidships.

0.5
1.
1.5
2.
2.5
3.

63

BALANCE OF BODY AND BALANCE OF SAIL.

Statical resistance of a thin plate not in point.

The mere statical resistance of a thin plate, floating vertically, to lateral motion, is collected at its centre of pressure, not at its actual centre of gravity. But there is less practical error in calculating by means of the latter, for several reasons; among which, one is, that when lateral motion has once begun, the water is heaped up in front of the plate, while a hollow is formed behind it. This creates a resistance at the surface, which more than compensates for the increased pressure of greater depths. In rapid motion, the centre of lateral resistance is found, in practice, to be considerably above the centre of gravity of the longitudinal section, instead of below it, as is the centre of hydrostatic pressure.

Effect of motion.

But the centre of lateral resistance of a ship with a full bow and water-line forward, is shifted forward from the moment she has speed, because the resistance on the lee bow is greater than on the weather bow, and because the resistance to a bow in motion is much greater than to the stern. The leeward motion also makes the resistance fall more directly on the bow than on the stern. Next, the lee bow is altered in form when the ship heels over by the wind, and becomes fuller than the weather bow. Hence such vessels, as they increase in speed, experience increasing pressure on the bow and not on the stern, thus driving the bow up into the wind, and allowing the stern to drift to leeward.

This disturbance of balance of the lateral resistance of the body of the ship has to be met in two ways. The ship has to be trimmed by the stern, which helps to bring back the centre of lateral resistance to the middle. Or it may be met in another way—more sail may be carried forward to counteract this effect.

Met by trimming sail.

The shift of centre of effort of all the sails forward, is the mode of correcting this disturbance of the centre of resistance, which is most employed by Naval Architects; but as it is not possible to make this adjustment absolute beforehand, each form of ship has its own peculiarity in this respect. One ship will balance her sail with its centre exactly on the middle of the water-line, another will carry it one-tenth of her length forward

Proper balance point for sail.

of the middle. As a general rule, any vessel having her bow water-lines convex, may be expected to carry her balance point of ship to balance point of sail—whether going free, or on the wind—one-twentieth part of her length before the middle, reckoned on the water-line, and nearly on an even keel. If the centre of the longitudinal vertical plane be made out of the middle of the length, the centre of effort must follow it.

CHAPTER XVIII.

OF THE PROPORTION, BALANCE, DIVISION, AND DISTRIBUTION OF SAIL.

A fine, fast frigate, in a ten knot breeze, can carry 36 square feet of sail for each square foot of area of midship section, and be the better for it; if she carried more, she might be pressed over so much, as to go slower, hence it has been common to provide a sail area of 36 square feet of canvas for one square foot of midship section; this proportion can be considerably exceeded by yachts and despatch vessels—even up to 100 square feet, but such vessels are mere sailing phenomena rather than ships; nevertheless, for light winds, all vessels carry a great quantity of light sail beyond their proportion of regular working sail. Proportion of sail to midship section usually 36:1.

Taking sail area in the proportion of 36 square feet of sail to one square foot of midship section, is merely saying how much canvas the ship should have, in order to drive her. Whether she will be able to stand up under it, and whether under it she will prove leewardly or weatherly, are other questions—questions of stability and balance of sail. All ships tend, under a side wind, to drift to leeward; the only preventive to this, is the extent of the immersed longitudinal section, which offers resistance throughout the whole of the length and depth of the ship in the water. The dimensions and shape of this section determine the arrangement and balance of the sail, and a ship should be sufficiently weatherly to carry an area of sail, fore and aft, not less than six times the area of this under-water longitudinal section.

As a first step to the consideration of the distribution and balance of sail, draw this section of the ship under its proper water-line, and copy it by drawing above it a similar section in the air, six times as high; this call the equivalent sail area, since it shows, without regard to the kind of ship, the quantity and disposition of sail which she may carry; and, in short, is what the sails might be, or would be, if she could conveniently carry them all in one. Equivalent sail area.

Indeed, a vessel with one sail is perhaps more effective than with any other number: but the larger the vessel, the more must her sails be sub-divided for convenience of handling. There is also a limit to the size, at which sails can be made strong enough and stretched flat. As one sail.

If the vessel were so small that she could carry the whole in one sail, she would be what is commonly termed "a lugger," and the sail "a lug sail." It is to be remarked, that the centre of effort of the wind on this sail, will be precisely over the centre of resistance of the longitudinal section in the water, and so there will be a perfect balance of sail.

If the vessel be too long to enable the sail to be carried in one, it may be carried in two, without much alteration in shape, and such a vessel will be the common lugger with two masts. As two or more sails.

OF THE PROPORTION, BALANCE, &c., OF SAIL.

In like manner the sail may be divided into three, and hung on three masts—then the vessel will be a three masted lugger.

Thus it is plain that this equivalent sail may be obtained indifferently by one, two or three sails, on one, two or three masts, as a matter of convenience merely, and that perfect freedom to make any decision as to distribution is given, provided only that the place and size of the sail, and therefore the balance, is maintained.

In light winds it may be desirable to carry additional sail; all that it is necessary to observe is, that the additional sails be so placed and proportioned as not to disturb the original balance.

By means of the bowsprit, the sail may be carried forward until it ends in a point, taking care, however, to extend the sail backward also sufficiently far to offset the addition in front, otherwise the wind will tend to make the vessel sheer round, and the balance will be destroyed. The whole sail will thus become one large triangle. This form is extremely convenient for vessels carrying fore and aft sail, but these additional sails, fore and aft, may be and indeed are, mere patches. They are used simply as balancing, or directing, or steering sails, to steady the vessel, without any regard to their propelling power. Such a triangular sail is sometimes carried by a single mast, and sometimes divided, in the same manner as lug sails, and it is curious to observe how differently the whims of sailors may be indulged as to the mode of supporting and carrying these sails; the single sail being equally well carried by an upright mast in the centre of the vessel, by a mast in the bow raking violently aft, and by a mast aft raking as extremely forward, the one condition being fulfilled of leaving the balance of sail unchanged.

Centre of effort of sail.
As all triangles on the same base, having the same height, have the same area, when once a triangular area of sail is obtained, it may be changed in shape at pleasure, provided the same height is maintained. It is to be observed, however, that as the shape of the triangle is changed, the place of effort of the sail is shifted with it. For instance, two triangular sails may have equal areas, but their centres of effort may be in different lines perpendicular to the base, owing to their change of shape.

The balance of sail will be destroyed, if, in dividing the sail area, care is not taken to see that in any new distribution made, the place of the centre of effort is not shifted by that distribution; not only must the portions cut off from one part of the sail area, be supplied in quantity by another, but care must be taken that in their new positions the new parts do not gain or lose power of balance; power of balance being effectual distance: the designer must therefore know how to calculate the exact effect of sails placed at different distances.

To calculate the balance of sail, there are two simple and convenient principles.

How found in triangular sails.
A triangular sail has its centre of effort in the line which joins any of its corners to the middle of the opposite side, and is nearer to the side than the corner, in the proportion of 1 to 2; so that by dividing the line into three equal parts, and marking the division which lies nearest the side, you mark the centre of

OF THE PROPORTION, BALANCE, &c., OF SAIL.

effort of the sail; or it may be found by drawing lines from two corners to the middle point of the opposite sides; where these lines intersect is the centre of effort.

Now it fortunately happens that the shapes of all sails, if not triangles, can be divided into triangles, merely by drawing a line through two opposite corners; each part of the sail can thus have its centre of effort separately found.

Having thus found the centres of effort of all the sails, or of their separate parts, the next question in order is: How find the joint effort or effect of any pair, or any number of triangular sails, or parts of sails? This is done by the principle of balance, which is as follows: *How compounded in sails of another shape or in two sails.*

In order that two equal sails may balance, they must be at equal distances from the point round which they are intended to balance, otherwise the one at the greatest will sway the other; hence equal sails will only balance at equal distances. The equal distances are to be reckoned from the centres of gravity of the respective sails; if, therefore, there are only two equal sails to the vessel, the balance is easy, for it is only necessary to place their centres equidistant from the balance point in the ship, and the sails will balance. The joint centre of effort of two equal sails, therefore, lies in the line of, and half way between their respective centres of effort.

But it may be a pair of unequal sails, and unequal in any proportion, say 2 to 3. The way to balance them is to give the small sail the longer end of the balance, and to give the longer end the same preponderance in length over the shorter, that the larger sail has over the smaller; thus the longer distance combined with the smaller area of sail, balances the larger area combined with the shorter distance. To see that they are equal, it is only necessary to multiply the areas by their respective distances from the centre, when these products are equal, the balance sought for is obtained.

Suppose, on this principle, it is required to find the centre of effort of a sail composed of two triangular parts. Find the centre of effort of each part, (by the method before given,) join these centres by a line, divide this whole line into as many equal parts as there are fathoms of area in the whole sail, give to the lesser portion a greater number of these parts and to the greater sail area a less number, dividing the line in the inverse proportion of the areas; the point of division is the joint centre of effort of the sail.

If now there were a number of sails, some on one side and some on the other of an intended balance point, and the question were asked, whether they balance: it would be necessary to multiply the areas of all the sails by their distance from the balance point, and, if the sum of the products on the one side were equal to the sum of the products on the other side, there would be a balance. To obtain a balance, therefore, it is only necessary to contrive that the sums of the products of the sail areas on opposite sides by their distances from the balance points (or their moments) shall be equal. *For a number of sails.*

OF THE PROPORTION, BALANCE, &c., OF SAIL.

How balance may be rectified. It is plain, therefore, that to bring about a balance where it does not exist, it is necessary either to substitute a larger sail on the wanting side for a smaller one, or to shift the place of sail nearer to or farther away from the centre, as required. A ship whose sails are ill balanced, may have the defect corrected in practice, by setting different quantities of forward or after sail, or the defect may be rectified on a larger scale by shifting the place of the masts, or in a smaller degree, by causing a mast to rake more or less forward or aft. Where these remedies may be inconvenient or impossible, the centre of resistance of the body of the ship may be shifted towards the centre of effort of the sail by trimming the vessel a little more forward or aft, as it is plain that trimming by the stern will bring it aft, and trimming by the head will bring it forward.

Place of masts. It is now necessary to establish proportions, according to which the masts and sails of a ship may be divided and distributed. Take for this purpose a vessel with three masts, and suppose her to be of the wave form—to be on an even keel, her length to be divided into ten equal parts, and her bowsprit to extend so as to bring the centre of the jib 5.41 of such tenth parts beyond the stem, the extremity of the spanker being one-tenth part beyond the stern. For the distribution of sail make the following division.

Distribution of sail on the masts. Divide the sail area into 24 equal parts: 7.071 of these are for the fore-mast; 10 for the main-mast; 1.65 for the spanker; 3.35 for the other sails on the mizen-mast; and the remainder, or 1.929 for the jib.

The place of the mizen-mast is one-tenth from the stern, of the fore-mast two-tenths from the bow, and of the main-mast three-tenths from the mizen-mast, or one-tenth from the middle, leaving four-tenths between the fore and main-masts.

Reckoning from the centre of lateral resistance of the vessel, which, on an even keel, is the middle of her length, we have the following arrangements:

Sail.	Quantities.		Effective distances.		Total effects or efforts.
Spanker,	1.65	×	5	=	8.25
Mizen,	3.35	×	4	=	13.40
Main,	10.	×	1	=	10.
					31.65 *after* moments.
Fore,	7.071	×	3	=	21.213
Jib,	1.929	×	5.41	=	10.436
Total,	24				31.649 *fore* moments.

The explanation of the above is simple; the five parts which form the sails on the mizen, consist of 3.35 of the square or upper sails, and the other 1.65 of the spanker, which spanker has its centre of effort one division farther aft than the upper sails: the parts which form the upper sails are therefore multiplied by 4, and give an effect of 13.40, while the parts which form the spanker, act at a distance of 5, and give an effect of 8.25.

The ten parts which form the main-mast sail area, are only at one division from the centre, and give an effect of ten; therefore the total effect of the after sails is represented by figure 31.65,

OF THE PROPORTION, BALANCE, &c., OF SAIL. 69

In the same manner the moment of the sails forward is found to be, as given in the table, 31.649.

This gives the balance of sail required, and it may be observed that the jib, though small, has as much absolute effect on the balance of sail, as all the sail on the main-mast, nay, rather larger, while the mizen, though comparatively small, actually balances the fore-mast. It is convenient to remember, in working a ship, that the sails on the main-mast and the jib balance each other—as also the sails on the mizen and the sails on the fore-mast—either set alone could work the ship. It may also be observed, that the jib may be made to balance exactly the upper sails on the mizen-mast. *[margin: Jib balances the sail on main-mast. Sails on the fore, balance the after sails.]*

It will thus be seen that the total moments of the sails forward, are represented by 31.649—the total moment of those abaft by 31.65, and the total number of component parts of sail are represented by 24.

It now remains to be considered, how to proportion the various sails on these different masts.* *[margin: Proportions for sails.]*

1st. Of the 5 parts of sail on the mizen, 1.65 go to the spanker, and the remainder is divided between topsail, top-gallant sail and royal in the following proportion: Mizen topsail, 1.518, mizen top-gallant sail 1.073, and mizen royal 0.759.

2nd. Of the ten parts of sail which go to the main-mast, 3.3 form the course, 3.035 the topsail, 2.147 top-gall't sail, and 1.518 the royal.

3d. Of the 7.071 which go to the fore-mast, 2.3333 form the course, 2.147 the topsail, 1.518 the top-gall't sail, 1.073 the royal.

Or condensed in tabular form as follows:

Masts.	Sails.						
Mizen.	Spanker,	1.65 ×	5	=	8.25	21.65	
	Topsail,	1.518 ×	4	=	6.072		
	Top-gallant sail,	1.073 ×	4	=	4.292		
	Royal,	0.759 ×	4	=	3.036		31.65
Main.	Course,	3.3 ×	1	=	3.3	10	
	Topsail,	3.035 ×	1	=	3.035		
	Top-gallant sail,	2.147 ×	1	=	2.147		
	Royal,	1.518 ×	1	=	1.518		
Fore.	Course,	2.333 ×	3	=	6.999	21.213	31.649
	Topsail,	2.147 ×	3	=	6.441		
	Top-gallant sail,	1.518 ×	3	=	4.554		
	Royal,	1.073 ×	3	=	3.219		
	Jib,	1.929 ×	5.41	=	10.436		

But there is a third question to solve. The proportion of sail on each mast has been obtained—the proportion of area of each sail has been obtained—there remains to be found the proportionate dimensions of each mast, which may enable them to carry their respective sails. *[margin: Proportions for masts.]*

In a three masted ship, it is necessary, both for symmetry of appearance and for balance of sail, that the proportion of sail on each mast should be tolerably similar; for example—on the

---
* These proportions, it must be observed, are for vessels constructed on the "wave" principle.

OF THE PROPORTION, BALANCE, &c., OF SAIL.

largest mast should be the largest topsail; on the smallest mast the smallest topsail, and so on. It is also necessary that the sizes of masts and spars should bear a due proportion to each other throughout.

The following proportion of sails will accomplish all this. Taking the three masts to have four sails, all similar, then from the proportion before given, namely:

Areas of sail—Mizen 5   Fore 7.071   Main 10.
Being in the proportion of 1   1.4142   2.

In order to make up this proportion, it is only necessary that all the sails on the three masts should be in, as nearly as possible, the following proportion.

 5   7.071   10

The sails on all the masts will have the proportion required,

 1   1.4142   2

Proportions for yards. For example, when the cross-jack yard has for its breadth of sail 50 feet, then the fore yard 70.71 feet, main yard 100 feet, or in that proportion.

The corresponding topsail yards should be in the same proportion, namely:

Mizen topsail yd. 35.35   Fore 50   Main 70.71
Miz. T.-gall't yds. 25.   " 35.35   50
" Royal "   17.67   " 25   35.35

It is obvious, also, that the lengths of those parts of the masts and spars which carry sail, should bear to one another the similar ratio of

Mizen 5   Fore 7.071   Main 10
Or,   1   1.4142   2

Specific example. With these general proportions in view, proceed to complete the arrangement of sail on a given ship, say of 550 tons burthen, whose length on the water-line is 150 feet, and draft on an even **Total sail area.** keel 16 feet 8 inches. Taking six times the draft of water, or 100 feet, this gives the height of the equivalent sail area 100 feet; which, by a length of 150 feet, gives a total sail area of

$150 \times 100 = 15{,}000$ feet area.

To place the masts. First—to place the masts, divide the length of water-line into 10 equal parts.

Distance of the mizen-mast from aft, $= .1$ of $150 = 15$ feet
    "   "   "   fore-mast from forward, $= .2$ " $150 = 30$ "
    "   "   "   main-mast from mizzen, $= .3$ " $150 = 45$ "
    "   "   "   main-mast from fore-mast, $=.4$ " $150 = 60$ "
                                                      ———
                                                      150 "

Sail area on each mast. Second—to proportion the sail area on each mast.

Mizen, five twenty-fourths of 15.000 $=$ $5 \times 625 = 3125.$
Fore,   -   -   -   - $= 7.071 \times 625 = 4419.375$
Main,   -   -   -   - $= 10 \times 625 = 6250.$
Jib,   -   -   -   - $= 1.929 \times 625 = 1205.625$
                                    ————
                                    15.000

OF THE PROPORTION, BALANCE, &c., OF SAIL.

Third.—To proportion the sails on each mast.

Mizen.
$\begin{cases} \text{Spanker,} & 1.65 \times 625 = 1031.25 \\ \text{Topsail,} & 1.518 \times 625 = 948.75 \\ \text{Top-gall't sail,} & 1.073 \times 625 = 670.625 \\ \text{Royal,} & 0.759 \times 625 = 474.375 \end{cases}$ 3125.00

Fore.
$\begin{cases} \text{Course,} & 2.303 \times 625 = 1458.125 \\ \text{Topsail,} & 2.147 \times 625 = 1341.875 \\ \text{Top-gall't sail,} & 1.518 \times 625 = 948.75 \\ \text{Royal,} & 1.073 \times 625 = 670.625 \end{cases}$ 4419.375 } 15.000

Main.
$\begin{cases} \text{Course,} & 3.3 \times 625 = 2062.5 \\ \text{Topsail,} & 3.035 \times 625 = 1896.875 \\ \text{Top-gall't sail,} & 2.147 \times 625 = 1341.875 \\ \text{Royal,} & 1.518 \times 625 = 948.75 \end{cases}$ 6250.00

Jib, = 1205.625

Area of each sail.

Or in other words,

	Spanker.	Fore-course.	Main-course.
	1.65	2.333	3.3
Or in the proportion of	1.	1.4142	2.
	Mizen Topsail.	*Fore Topsail.*	*Main Topsail.*
	1.518	2.147	3.035
Or in the proportion of	1.	1.4142	2.
	Mizen Top-gall't sail.	*Fore Top-gall't sail.*	*Main Top-gall't sail.*
	1.073	1.518	2.147
Or in the proportion of	1.	1.4142	2.
	Mizen Royal.	*Fore Royal.*	*Main Royal.*
	0.759	1.073	1.518
Or in the proportion of	1.	1.4142	2.

Again,

	Topsail.	Top-gall't sail.	Royal.
Mizen.	1.518	1.073	0.759
Or in the proportion of 2.		1.4142	1.
Main.	3.035	2.147	1.518
Or in the proportion of 2.		1.4142	1.
Fore.	2.147	1.518	1.073
Or in the proportion of 2.		1.4142	1.

Now to get the length of the yards: The lower yards are at once found by taking the square root of twice the area of the courses, and for the mizen-mast, as the area of the spanker is equal to the same area as that of a course, if there had been one for the cross-jack yard, twice the square root of the area of the spanker must be taken.

Length of yards.

From this is found,
The length of the Main yard 64.23 or in the proportion of 1.
" " " Fore yard 54. " " " 0.8409
" " " Cross-jack yard 45.41 " " " 0.7071

These proportions give at once the length of yards and hoist of sail; for multiplying the length of the main yard by 0.8409, we get the length of the fore yard and main topsail yard; by multiplying the length of the fore yard or main topsail yard by 0.8409, we get the length of the cross-jack yard, fore topsail yard and main top-gall't yard, and so on, always excluding yard arms. These same proportions answer for the hoist of sail; or in other

Hoist of sails.

words, half the length of the main yard is the hoist of the main topsail; half the length of the main topsail yard is the hoist of the main top-gall't sail; half the length of the main top-gall't yard is the hoist of the main royal

Summary.

It will thus be seen, that by this arrangement there is one yard of the length of the main yard, two yards of the length of the main topsail yard, three yards of the length of the main top-gall't yard, three yards of the length of the main royal yard, two yards of the length of the fore royal yard, and one yard of the length of the mizen royal yard. Working the above quantities out for the vessel whose area of sails have been calculated, is obtained for the length of

Main.   *Fore.*   *Mizen.*

Main yard   = 64.23   Fore yard   = 54.   Cross-jack yd. 45.41
Topsail yd.   = 54.   Topsail yd. = 45.41   Topsail yard 38.18
Top-gall't yd. = 45.41   Top-gall't yd.= 38.18   Top-gall't yd. 32.41
Royal yard   = 38.18   Royal yard = 32.41   Royal yard   27.26

And when, with these figures, the areas of the different upper sails are calculated, it will be found that the quantities found in this manner, and the quantities found in the first manner, will agree with great precision.

It will be seen, that with the foregoing arrangement, a perfect balance is obtained; that is, the centre of sail falls exactly in the same perpendicular with the centre of lateral resistance. Now in some ships, it is preferable to have the centre of effort some distance forward of the centre of lateral resistance. This is easily accomplished by means of the jib.

How the jib may be used to alter balance of sail.

The centre of the jib in the foregoing calculations, was situated at 5.41 from the centre or middle division. Now by merely shifting its centre to six divisions from this middle, which in the vessel of 150 ft. length, would be 90 ft., the centre of effort would be brought forward 5 feet. This may seem difficult to do, as the masts are fixed and the jib stay cannot be shifted; it therefore remains to alter the shape of the jib, this is done in the following manner:

Erect a perpendicular line on the sixth division from the middle of the water-line, then on this perpendicular the centre of the jib will be situated. Lengthen this perpendicular until it meets the jib stay, then lay off from this intersection equal distances, up and down along the stay as far as convenient, the sum of these two distances will form one side of the jib. The area of the jib being given, divide this area by one-half the side, and with the quotient as length, draw a line parallel to the jib-stay until it intersects the perpendicular line, join this point with the two extremities on the jib-stay and there is obtained a shape of jib of the given area, and with its centre falling exactly at the sixth division from the middle, or one-tenth of the length beyond the stem. But this degree of accuracy is much greater than is required for practice, and it is necessary to guard against the attempt to fix these points in the design too closely before taking into consideration a multitude of practical points of convenience, use and taste, which go to regulate the dimensions of sails. In the first place, it must never be forgotten that nearly all ships carry weather-helm, and that this proportion of weather-helm generally increases with the wind.

It is to be observed, that the design of the sails having been made in proper balance, any change made to correct defects in the form of the body, should not be allowed to derange either the proportions or places of the sails; but, for this purpose, the whole of the sails should be removed to their new place, and not shifted with respect to each other, unless due regard be had to maintaining their balance. *If alterations in the ship are made, balance must be maintained.*

Another point for consideration is, that if masts are made to rake, instead of standing upright, it must not be forgotten that rake may shift the relative distance of the sails.

A further point is, that the convenience of the ship herself may interfere with the disposition of sails. A high forecastle will shorten the foot of the foresail. A poop may seriously interfere with the spanker. These are points which must on no account be neglected. *Convenience of ship sometimes dictates sail to be used.*

Perhaps the most important point that can be kept in view in the study of the balance of sail, balance of body, placing of masts, proportion of spars, and sub-division of sails, is this, that in all circumstances the ship should be able to carry the greatest quantity of sail with the least possible action of rudder. In a perfect wave form, perfectly balanced, this has been done, and in a fast sailing clipper it is vital. In such a vessel, the whole of the sails mentioned would be carried, whether the wind was light or fresh, without retarding the ship by the action of the helm. When it came on to blow hard, it would only be necessary to furl the three top-gall't sails, and the rest of the sails would remain in perfect balance; blowing harder, the topsails might all be reefed and a balance still maintained; blowing a gale, the spanker, jib, foresail and mainsail might be taken in, and yet a perfect balance exist under close reefed topsails and storm jib. Thus, in a ship built on the wave line theory, even in heavy weather, the Captain would find his ship handy, fast and under perfect command; but if the vessel were not a wave vessel, the following changes would take place: As soon as it came on to blow fresh, the spanker, which is a most powerful sail, would be found to cause an excessive degree of weather-helm, and would have to be taken in, but that would spoil the balance and the jib would follow the spanker, giving place to the topmast staysail, which would at once reduce very seriously the way of the vessel, and it would be want of balance and not stress of weather which did it. If it came on to blow hard, it would soon be necessary to take all sail off the mizen, except perhaps a small storm sail for lying-to. *Importance of balance of sail.*

Balance of sail perfect only in wave form.

In ships of this class, nothing but experience will tell under what sails the ship will balance, and what she will not carry; but one thing is certain, that in light winds and strong ones, the balance will be entirely different, which is not the case in the wave formed ships.

OF THE PROPORTION, BALANCE, &c., OF SAIL.

TABLE OF AREAS AND POWERS OF SAILS WITH FOUR YARDS ON EACH MAST.

Six times the area of the immersed longitudinal section is taken as the standard sail area of the ship, of which, for—

The area of all the square sails on Mainmast = 10·24 of the whole area.
" " " " " " Foremast = 7·071·24 " " "
" " " " " " Mizenmast = 3·35·24 " " "
" " " " " " Spanker = 1·65·24 " " "
" " " " " " Jib = 1·929·24 " " "

} 24·24

The powers of these qualities are:
For Main area 1
" Fore " 3
" Mizen " 4
" Spanker " 5
" Jib " 5·41

Or,

The area of all the square sails on Mainmast = 0·4167 of the whole area.
" " " " " " Foremast = 0·2946 " " "
" " " " " " Mizenmast = 0·1396 " " "
" " " " the Spanker = 0·0687 " " "
" " " " the Jib = 0·0804 " " "

The moments from the middle of the water-line are:

Spanker.
1·65
× 5
=
8·25

Mizen.
3·35
× 4
=
13·40

Main.
10
× 1
=
10
13·40
8·25

After moments 31·65 =

Fore.
7·071
× 3
=
21·213
10·436

31·649 Fore moments.

Jib.
1·929
× 5·41
=
10·436

OF THE PROPORTION, BALANCE, &c., OF SAIL.

The area of the Main Course   is = 0.1375 of the whole area
"   "   "   Topsail   " = 0.1265   "
"   "   "   Top-gall't sail " = 0.0895   "
"   "   "   Royal   " = 0.0632   "
"   "   Fore Course   " = 0.0972   "
"   "   "   Topsail   " = 0.0895   "
"   "   "   Top-gall't sail " = 0.0632   "
"   "   "   Royal   " = 0.0447   "   = 1
"   "   Mizen Topsail   " = 0.0632   "
"   "   "   Top-gall't sail " = 0.0447   "
"   "   "   Royal   " = 0.0317   "
"   "   "   Spanker   " = 0.0687   "
"   "   "   Jib   " = 0.0804   "

PLACE OF THE MASTS.

The length of the ship on the load water-line from the after part of the stern-post to front of the stem, is divided into ten equal parts:

The Mizen mast is placed at one-tenth   from the stern-post.
"   Main "   "   "   " four-tenths "   "   "   "
"   Fore "   "   "   " eight-tenths "   "   "   "

CO-EFFICIENTS OF PROPORTIONS OF SPARS AND SAILS.

With 4 yards on each mast the Course   is = 0.33   of whole sail area of mast   The area of the Course being given
"   "   "   "   "   "   Topsail   " = 0.3035   do.   the length of the yard is equal to
"   "   "   "   "   "   Top-gall't sail " = 0.2147   do. $\sqrt{2 \times 0.1375 \times A} =$
"   "   "   "   "   "   Royal   " = 0.1518   do. $\sqrt{0.2750 \times \text{area of ships sails.}}$

CHAPTER XIX.

OF SYMMETRY, FASHION AND HANDINESS OF SAIL.

Fashion of sail a question of seamanship. Hitherto the sails have been studied with reference to their effect on the ship, in so far as concerns the work of the Naval Architect. Whether they are well proportioned in size to the ship; whether they are well balanced so as to leave the ship free in her movements; whether they are so proportioned in dimension, that they drive without overpowering the ship; whether they can be varied in quantity to any extent, without derangement of balance, and always leave the ship under command of the helm; when these questions are satisfactorily answered, then the first great requisites of the Naval Constructor are accomplished. But other things are demanded, besides this first essential—their use to the ship. The seaman must be satisfied with the figure, distribution and cut of his sails—besides this, they must suit his convenience and use. They must set well, stand well, draw well, be easily set, easily worked, easily reefed, easily taken in; in short, be conveniently, easily and safely handled. On this point, the will of the seaman should rule the design.

The quantity and balance of sail, is the business of the Naval Architect; the symmetry, fashion and cut of the sails, is the vocation of the seaman, not of the landsman. The Naval Architect has now to consider how he shall give the seaman all he wishes, in regard to fashion and symmetry, without compromising the other conditions on which the ship is designed. This requires skill—but for this purpose, all that has been said about balance of sail and of ship, forms an excellent basis, on which may be grafted any amount of fashion and of fancy, of fitness and of handiness.

Suppose a full rigged ship to have been designed; and the place of all the principal sails, their areas, and their dimensions to be laid down on a sail-draft, by the rules already given; the question now raised is, How may the fashion of the sails be altered, without disturbing their balance, or changing their quantity?

There is manifestly a great variety in fashion for the same area.

For every square sail, therefore, there arises three main questions:

1st. Taper of sail, or diminution of the head of each sail, compared with the width at the foot.

2d. Proportion of height to width, or spread of sail in proportion to hoist,

3d. Sub-division of sails on a mast.

Taper of sails. I. *Diminution of the head of the sails on a given mast.*

The sails on the same mast may all have the same taper, diminishing in one straight line; or they may vary in taper,

SYMMETRY, FASHION AND HANDINESS OF SAIL.

It is obvious, that whatever reason exists for a certain taper in a given sail, will apply equally to that above it. That the sails on all three masts should have the same taper, one and all, seems evident. There seems to be a preference for having one proportion for the diminution of the head of the sail running through all the higher sails of the same mast, especially where the sub-divisions are numerous. But, on the other hand, it is a frequent practice to narrow most, the heads of the higher sails.

The argument for diminishing the heads of the sails, is that the higher masts and gear are lighter than the lower, and, therefore, less able to carry heavy and large sails and yards. On the other hand, is the argument, that the loftier sails are not spread in bad weather, but are taken in when it blows hard, so that being fair weather sails, they should be large, or else they are of but little use.

The latter consideration is entitled to considerable weight. The lofty sails should, it is thought, have a wider spread and a smaller proportion of height to width than has been usual hitherto. There is a growing tendency in fast vessels to carry large and low sails, and to obtain greater spread of sail with less hoist.*

Moreover, with the adoption of iron and steel as a material for masts, and wire rope for rigging, sails of great spread and moderate hoist, will it is believed, be more and more used.

Three things must be remembered, in considering what diminution of sail may be adopted in any given ship.

1st. That, by increasing the spread of the lower sails and tapering rapidly the upper, the centre of effort of the sails is lowered.

2d. That, by narrowing the upper sails, they become of less area and of less value.

3d. That, in altering the taper, it is only necessary to remember that, by so much as the alteration adds to one part of the sail area on a given mast, by so much also, shall it diminish the area at another part.

Thus, any amount of change or diminution may be given to the sails on each mast, without changing the balance of sail area on the whole ship.

II. *Proportion of height to width in a given square sail, is a matter of choice.* _{Spread and hoist of sail.}

It seems that in proportion as ships sail faster, and are built finer and longer, the separate sails are made broader and lower, their yards longer, and their hoist less. By giving squareness to a sail, not only is a larger quantity of low sail carried, but the sails stand flatter and better on a wind. On the contrary, there is this consideration, that yards of great length are costly and heavy—heavy to carry and heavy to work; and that, by merely increasing the hoist, the same yard may be made to carry much more canvas and do much more work.

This goes in favor of increased hoist; but it loses weight from this further consideration, that a square sail of great height,

---

* This is especially the case in wave-line ships.

does not stand well on a wind; and that a fast ship will sail faster on a wind with square and low sails, than with high and narrow ones.

The fact that yards of great length are heavy to carry and hard to work, will therefore be a good argument in favor of narrow and lofty sails for slow ships, for short voyages, and for ships with small crews. On the contrary, in long voyages, and with plenty of able seamen, spread being of value, long yards and moderate hoist are preferable.

The limits of proportion taken are these: When the hoist of a square sail is made equal to its spread, that is to be reckoned an extreme height of sail. When the hoist is one-half of the greatest width, that is to be reckoned a broad and low sail. They ought not be lower, since it seems wasteful, because a sail of that height will stand close to the wind; therefore, that is assumed as a standard proportion of height to width.

<small>Sub-division of sail on a mast.</small>

III. *Sub-division of sails on a mast.*

It is plain that the proportion of width to height of sail, may be considered apart from taper, or diminution of head; nevertheless, a rapid rate of diminution may better suit lofty sails, and a more gradual rate lower sails. But much of the symmetry of a suit of sails, depends on keeping some one proportion of height to width of sail, throughout the sails on the same mast, and throughout the sails on the different masts of the same ship.

In fast ships, there is a strong tendency in this direction; and it is believed, that the introduction of iron masts, frees the ship builder from the difficulty of finding spars of sufficient length and strength in the forests, and enables him to make masts of any length in one piece, without break or discontinuity, and this is a great encouragement to the adoption of symmetry and uniformity in the proportion and fashion of sail. It seems plain, that when some one proportion of height to width has been selected, as possessing the requisite qualities in the best practical degree, there can be no sufficient reason for adopting that proportion in one sail on a mast, and rejecting it in the others.

Take therefore, for example sake, the sails on one mast, and divide them, so that they may all have one proportion of spread and hoist. That sub-division may be altered in any way found most convenient for working. In men-of-war, the topsail is the great working sail of the ship; it is generally of great hoist, and may be taken as an extreme proportion. In the double topsail sailing clipper, the same sail is cut into two sails, often of a ridiculously small hoist. These are two extremes between which there should be some medium. It is thought not out of place to repeat here, that sub-division of sails is more a matter of seamanship, than of naval construction—is more, in fact, a question of working a ship, than of designing one. Generally speaking, the sails liked best will be worked best. What the seaman likes, will depend not merely on his experience, but on the power at his disposal to work his ship, and on the value that speed may have to him. Given—a stable, fine, fleet ship for long voyages, it is thought better to have sails not high, but of great spread. For short voyages, narrow seas, moderate speed,

SYMMETRY, FASHION AND HANDINESS OF SAIL.

and a small ship's company, narrow sails, lofty and easily worked, may be preferred: and in like manner, sails few and large, or many and small, have corresponding advantages or disadvantages.

Hitherto, reference has been made mainly to sails of a quadrangular shape or square sails, which are not only the most universal of form and arrangement, but are universally used on the largest scale. Triangular sails are not less valuable, but are to be reckoned in some sort as subsidiary sails. They take their form almost inevitably from other considerations, to which they are subordinate. Thus, even the jib of a man-of-war, the chief triangular sail, takes its shape and proportion almost exclusively from the angle of the jib stay, and is decided in shape by the proportion of masts and direction of rigging, which have been determined by precedent considerations. *Fore and aft sails.*

If there is less scope for choice and design in triangular sails than in square sails, there is this compensating virtue in the former, that they are accommodating enough to take any shape without loss of value. A jib covering a given length of its boom, is of the same area, provided it rise to a given height, measured square off the line of its boom, and does not vary with the steeve, and so long as it rises to the same height, its centre of effort will be at the same height taken square from the boom.

There is another point in which a triangular sail differs from a square sail—it draws well, independent of its height. So long as a triangular sail is not too wide fore and aft, it will set flat, close to the wind, and without the large belly which great height would give a square sail. The chief virtue of triangular sails is this special quality of setting flat, and going close to the wind.

Spankers and the sails also have the same advantages as triangular sails of standing well, and keeping flat, close to the wind, but the gaff has the disadvantage of tending to sway over to leeward; and the head of the sail shakes while the foot stands.

In calculating the balance and distribution of triangular sails, and fore and aft sails generally, it is a matter of indifference, what the sort of sail is; a fore and aft sail may be substituted at any point for a square sail, provided the same area is kept, and the balance point or centre of effort of the sail in the same place. The sails will balance the ship equally well, whether square or fore and aft.

But there is this radical difference between fore and aft sails and square sails. Fore and aft sails shift their centres of effort with their trim—they travel in circles round a fixed point, and they carry their centres round with them. Square sails never shift their centres of effort, so long as they are set flat; the centres are fixed points on the mast around which they turn. *Fore and aft sails shift their centres—square sails do not.*

This shifting of the centre in fore and aft sails, is of considerable importance, because it carries the centre of effort further forward as the ship's course goes off from the wind. It returns when close-hauled; but it must be kept in mind, that it is always a little forward of its calculated place, and this is perhaps one of the reasons why, in fore and aft rigged vessels, the centre of effort of the sails requires a smaller shift forward of the middle,

Fore and aft sails. in order to meet the shift of the centre of resistance of the body of the vessel as the speed increases. It must always remain a great point in favor of the square rigged vessels, that their sails pivot round their centres of effort, and keep their balance in every trim. On the other hand, it is a quality of the fore and aft rig to lie closer to the wind, and probably to yield a given sail area with a smaller quantity of top hamper, thus suiting well the chief purpose of modern sails, to serve as auxiliaries to the power of steam.*

But iron masts, spars and wire rigging, are daily coming more and more into use, and will eventually open up a new field for enterprise—at least in the merchant service.

---

*The majority of the French iron clads carry large fore and aft sails, in lieu of square sails—probably on this account.

CHAPTER XX.

CONDITIONS OF THE PROBLEM OF NAVAL ARCHITECTURE.

The professional duty of the Naval Architect is to frame and complete the design of a ship—the word design implying plan, use, or purpose; and, therefore, the first duty of the Architect is to ascertain accurately, note exactly, and conceive clearly, the intention and purpose which the vessel is designed to fulfill. *Duty of the Architect.*

If the case under consideration is that of a merchant vessel, to the owner then, the Naval Architect must apply for a clear understanding of all that the ship is meant to be and to do; and therefore the following questions may be of service in eliciting the information necessary before commencing the design of the vessel:

The owner must be asked—first, what he wants his ship to do? He may answer: To trade between New York and New Orleans. *Example.*

2. What kind of trade he proposes to carry on? Answer.—A miscellaneous trade, partly merchandise, partly passengers.

3. What quantity, bulk, and nature of cargo? Ans.—500 tons of dead weight; 25,000 cubic feet of bulk, for cargo in the hold.

4. What kind and number of passengers? Ans.—25 first class, 20 steerage passengers.

5. What sort of voyage? Ans.—Once a month, stopping nowhere on the way.

6. At what speed? Ans.—An average of 8 knots.

7. Carrying much canvas or little? Ans.—To depend mainly on steam, the sails being auxiliary.

8. At what estimated cost per voyage? Ans.—$1.75 per mile.

9. How much is the owner prepared to pay for his vessel? Ans.—$125,000.

10. How much is the owner prepared to pay for a more or less durable ship? how much for more or less durable engines and boilers? and how much for a more or less complete equipment? Ans.—Ship to be classed twelve years, A No. 1; engines and boilers to be those least likely to fail when wanted, most economical in repairs and consumption of fuel; and 15 per cent. preference to be allowed on the price of good engines and boilers over indifferent.

11. What draft of water? Ans.—Load draft not to exceed 15 feet; no other limit as to dimensions.

12. What class of Commanders and Engineers to be employed? Ans.—The best Captain and Engineer without reference to salary. (The owner will do well to select his Captain and Engineer and put them in communication with the Naval Architect before the ship is built.)

13. Is the ship to be confined exclusively to this trade or may she have in future to be employed on other voyages?

11

Now from the Captain and Engineer, the Architect may receive information on the following questions:

14. What is the true length of the voyage according to the course usually followed?

15. What has been the average performance of any known vessels on the line?

16. What would require to be the maximum speed of a vessel in good sailing trim, in order to realize an average working speed of eight knots an hour on the voyage?

17. What sort of ships and engines have hitherto been employed to do this sort of work?

18. With how many officers and hands as crew, and how many in the engine room, is this ship proposed to be worked?

19. Besides the room required for cargo, for passengers and for attendants, how much is to be reserved for machinery, for coals, for ship's company, for ship's stores, for provisions and equipment?

20. What is the exact nature of the equipment required for this peculiar voyage?

21. What are the weights to be carried under these respective heads?

Points of construction. These are the conditions of the problem, without which, as preliminaries, the design of the ship cannot even be begun, and all of them must be sought, and given to the Naval Architect at the outset, in order to prevent much of his work being mere waste.

Preliminary conditions. The result of all these inquiries will lead him to this most important and primary issue, which may be said to determine the chief characteristic of his ship, namely: the burden she must carry and the bulk she must stow. In addition to her own powers to swim, she must have power to carry, and the total weight *Bulk of weight.* she must carry when full, is 1000 tons. But the vessel herself will weigh a known quantity, a quantity either suggested to him by some vessel he already knows, or which he must find out by calculation; but suppose it be assumed that his ship will weigh 500 tons in addition to the 1000 tons before stated.

The ship, therefore, with her equipment, her freight and her stores, gives a dead weight to be dealt with in the design of 1500 tons. This is technically called "the total deep-load displacement of the ship," and forms the first condition of the problem. It is the dead weight to be carried, and the question is, how best to carry it? This is treated of, under the head of Displacement. (Pages 14 and 15.)

Peace and war. The foregoing, drawn from the necessities of the merchant service, will serve also to suggest a similar series of requisitions to be made before commencing the design of a vessel of war. The nature of the service on which a man-of-war is to be employed, the harbors she is to enter, the length of a voyage on which she may be sent, the number of her crew, the weight of her guns, ammunition, equipment and stores, and for a steam vessel, the power required to drive her at a given speed, and the coal required to take her a given distance, with a multitude of

PROBLEM OF NAVAL ARCHITECTURE. 83

particulars quite as minute as those given in the case of the merchant vessel, must be obtained by the Naval Constructor before he can commence his design.

A Naval Architect, for his own reputation, should refuse peremptorily to commence his design without full, specific and detailed conditions first obtained, either given to him by authority, or prescribed by himself if he have the power.

It is much too common a practice to ask a designer to build a ship-of-war, and to tell him that it will be time enough to consider all the details of her armament, equipment, special construction, and destination, after the design has been completed, and while the ship is in progress. This is a fallacy, it will not be time enough, it will be too late.

Most of the wretched failures in this country, have been produced by building the ships first, and settling what they were to do afterwards. The Naval Architect who respects his profession should refuse to design his ship until all the requisite data has been given him. Without this there can be no science of Naval Architecture, and no plan of a ship worthy of being called a design.

But, when these have been obtained, the Constructor should arrange them, reconcile them, and finally determine them, by setting them out in a formal manner, in what may be called the

SCHEME OF CONDITIONS OF CONSTRUCTION, *Scheme of conditions.*

which forms, afterwards, a programme of work to be done in forming the design of a ship.

Scheme for the Construction of a Merchant Steamer.

	Bulks. Cubic feet.	Weights. Tons.
A miscellaneous cargo,	25,000	500
Passengers—25 first class,	6,250	
" —20 second class,	3,000	
Engines and boilers (with water),	7,500	150
Fuel and Engineer's stores,	10,000	200
Equipment and sea stores.	7,500	150
Ship's hull and internal fittings,	17,500	350
Provisions and water,	2,500	50
Officers, engineers, servants and crew,	7,500	10
Spare capacity and weight,	3,250	90
Gross capacity and weight,	90,000	1,500

Voyage of 1500 sea miles, (knots)
A mean speed of 8 knots.
Load draft, 15 feet.
Speed in smooth water, 10 knots,
Fuel per mile, 1 1-2 cwt. (168 lbs.)

Ship's company: { Officers, 5 }
{ Engineer and assistants, 3 } 30 hands.
{ Crew and coal heavers, 22 }

Time of single voyage, 8 days.

Scheme for a Man-of-war Screw Steamer.

                                                              Bulks.   Weights.
                                             Cubic feet.   Tons.

Engines, boilers (with water.)
Engineer's stores.
Fuel.
Guns.
Powder, and tanks including space for
    light rooms.
Shot and shell.
Ordnance stores.
Water for 4 weeks, for ——— men.
Bread for 6 months, for ——— men,
Other provisions for 6 months.
Masts, yards, rigging and sails.
Spare sails and Sailmaker's stores,
Navigator's stores.
Boatswain's stores.
Carpenter's stores.
Boats.
Chain cables,
Anchors.
Officers' stores,
Paymaster and Marines' stores,
Galley and condensers.
Officers, crew and effects, (———)
Shaft alley.
Wing passages.
Ventilating passages.
Masts and hatchways.
Spare bulk and weight,
Weight of ship's hull,

CHAPTER XXI.

HOW TO DESIGN THE LINES OF A SHIP.

The easiest problem, which can ever be submitted to a constructor, will be taken as an example. Problem stated.

Suppose a case, which very frequently occurs in practice, that a certain length of ship is to be built—a certain breadth is given—a certain load draft of water, and a certain light draft of water, and that these are about the ordinary proportions of a ship; that no particular weight is to be carried, or work to be done, beyond sailing well, or steaming at a moderate speed, and that the purpose to be served is a fair, common mercantile trade, such as ordinary vessels will moderately well perform; of course the owner will expect, what he may with reason expect from a man of science and skill, that the vessel will be somewhat faster, easier, safer and more economical, therefore somewhat more valuable, than a vessel built, without design or calculation, by an unskilled man. This is a very ordinary task for a Naval Architect.

There are two ways in which he may set about building his vessel: he may either take the model of the vessel which is already the best that has been applied to the trade in question, and improve upon her; or he may at once throw all precedent overboard, and give his employer an entirely new design. The undertaking then will speedily shape itself as follows: Extreme length and extreme breadth being given, he may determine a midship section, such as will give him the requisite carrying power, with good sea going qualities. Next, he will determine a water-line which will give the highest speed and least resistance, of which that length admits; or he may decide to fit her for a given speed only, and adopt a water-line of greater capacity fit for that slower speed. Thirdly, he will adopt a convenient form of deck for the use and navigation of the ship, and on these principal points he will fill in, what may be called "a skeleton design," and frame an approximate calculation of the qualities of the ship, which may also be called "the skeleton calculation."

(1.) *To construct the midship section.* Water-line not arbitrary.

In the choice of the *midship section*, the Naval Architect is left free, to exercise with the greatest liberty his own judgment. *In the water-line, he has little or no choice:* Nature has fixed that for him. If he meddle with it, he displays ignorance or presumption, and the due punishment is a spoilt water-line; but the midship section he may vary as much as he wishes to. He may give the ship every sort of quality, by choosing it ill or well; and with a given water-line, he may produce all sorts of ships.

To illustrate this latitude of choice, suppose the architect to take three styles of midship section; and further, that it is necessary for each of them to have the greatest speed the length will allow. Choice of midship section.

The first of these sections is to carry extremely little cargo, to have little room, but to go as fast as she can be made to go with all the sail and steam power she can carry. These are the practical conditions of the yacht—the swift cruizer—the opium trader—or the privateer. What such a vessel requires can be readily contrived—for the conditions given, *make* the midship section, and leave not much to choice.

Such a vessel must be all shoulder and keel, and nothing else; she is like a race horse, lanky and leggy: by being all shoulder, with very little under-water body to carry, she will possess the maximum of power with the minimum of weight—her fault will be, that she must have an enormous keel—to prevent her from going to leeward; and this great mass of dead-wood or solid keel, exposes a large surface to the adhesion and friction of the water. Nevertheless, it is the form of greatest power with the least weight. The bottom of this midship section may be formed in two ways—it may either be made elliptical, to have a minimum of skin for adhesion, and be reconciled to this deep keel, by two hollow curves; or it may be reconciled to the keel by a long wedge bottom. The elliptical bottom is thought best for iron ships; but the other or peg-top shape has been much

For medium qualities. used in wooden ones. Next, suppose that the capacity thus got, is too small for carrying remunerative cargo, and that a cargo hold of a capacity more usual in mercantile vessels is required; in that case keep the same shoulders, and give a larger under-water body.

For weight. Now take a third design. The ship is to carry as much as is not inconsistent with good sea-going qualities; and she is to have room also for boilers and machinery of considerable power. This requires her sides to be nearly upright, her bottom *dead flat* amidships, with only so much off her bilges, as will not be inconsistent with what she is to receive inside This the form and arrangement of her boilers and machinery will generally determine, and the boilers and machinery in such a vessel should be treated *as ballast and kept low*.

Comparison of the three forms. In regard to these three midship sections, it is to be noticed, that they are prescribed in some measure by the uses of the ship; but the forms mentioned, come entirely from the judgment of the constructor, and whether they have been wisely or injudiciously selected, must be judged of after the calculations have been made of the various qualities, to which they give rise.

But there are one or two points, which occur to a constructor, at the first glance at these forms of under-water body. It is plain, the first is easiest, and the last hardest to drive. It would require much more sail (or power) to drive the two last than the first.

And it is equally plain, that the first is much better able to carry sail than the last. The area of midship section of under-water body is the thing to be driven; the area of the sail is the driving power: but the power of the shoulders to carry the sail upright, limits the quantity of sail the ship can carry. The bulk of the under-water body brings with it two evils—resistance to being driven through the water, and under-water buoyancy tending to upset the ship.

HOW TO DESIGN THE LINES OF A SHIP. 87

It is plain, from these considerations, that the first shape is suited for a fast ship under sail alone—the last is suited for a fast ship under steam alone, and the second form may do for a moderate quantity of both, or what is known as "the mixed system."

Of these three vessels, the following will probably be an approximation to their qualities.

The first will be powerful, weatherly, lively and fast. The last will be tender, easy, sluggish and roomy. By a proper medium may be obtained, in the second vessel, any compromise of these qualities which may be fancied.

Nothing has yet been said about the parts of midship section above water; but it will be noticed, that these grow naturally out of the form adopted under-water; and it will be observed, that the above water body should be proportioned to the under-water body. The object of this, is to give adequate lifting power in a seaway, in proportion to the heavier under-water body.

In the three designs, the midship section (a technical term) is far from being actually amidships, being placed at the point of greatest breadth, or nearer the stern than the bow, in the proportion of 4 to 6.

(2.) *To construct the chief water-line.*

For the side view of a ship, or vertical section, draw a horizontal line, representing her whole length at the main water-line, and erect at the end of it two perpendiculars. This line is called "the length of the ship, the length between perpendiculars, or the construction length." These perpendiculars are to be ruling elements of construction, and are called "*the* perpendiculars." Construction length.

Divide the length between them into ten equal parts; take four of these abaft for the length of the "*run*," and six of these forward for the length of the "*entrance*." Describe a semi-circle on each half for the chief breadth. Divide the length of entrance and this semi-circle into any (and the same) number of equal parts; the distance of the water-line from the centre line, opposite each division in length of the entrance, will be the distance of each corresponding division of the semi-circle, from the same centre line, and a line through all the points thus found, will be the true wave water-line of the bow. The water-line of the run will be different from this in one respect only—it will be necessary to draw parallel lines to the centre line, through each of these divisions of the semi-circle, which it may be convenient to call "the semi-circle of construction." On each of these lines, points may be found, just as if it were a bow line. These lines must be prolonged aft, beyond the points thus found, to a distance equal to that part of the line intercepted between the semi-circle of construction and the main breadth. The last found points in the parallel lines, give the line of main breadth. Wave-line entrance, how described.

Wave stern.

The chief water-line of the bow and of the stern, or the lines of entrance, and of run for the greatest speed which the given length will admit, are thus formed. The lines thus given are absolute, and will admit of no deviation, without some loss. Modification of water-line for stem and stern-post.

Nevertheless, some modification in the application of these lines, may be admitted as expedient, and one of them is obvious.

It will be seen, that the point of the bow is so extremely sharp, that it would be in continual danger of cutting everything that it touched, and being so fine, would run risk of being crushed by rough usage. The stem also, for the rough work of a ship, must be of considerable thickness, and the practical question is: How can the line at the bow be altered to get this thickness? Shall the fine part be cut off, and thus shorten the vessel? If so, she will be too short for the length determined on.

The answer to the question is: Draw the bow to a few feet longer than it is intended the ship shall be—then cut off this water-line to the length it is intended to keep. The result obtained, is a thickness of a few inches, remaining between the two sides of the water-line, which is just thick enough for the materials of the stem. By this means, is obtained the extreme length required, and also the strength of stem necessary for durability, and it will also be observed, that the bow thus gained, is of slightly greater capacity than the first attenuated line. This is, therefore, called "the corrected water-line of the bow." No such adjustment is necessary at the stern. It is enough there to insert the stern-post, simply by increasing the breadth between the lines, to admit the thickness of the stern-post—a deviation insufficient to cause a sensible difference in the performance of the ship.

For sheer or trim. There is another modification of these water-lines of the fore and after bodies, which may require further consideration. Both entrance and run have been put in the same plane, with the intention of leaving them so in the construction of the ship; but it may be necessary afterwards for some good reason to change it.

The Constructor must be prepared to do so. The main water-line may require to go lower or higher in the vessel than is now proposed, and the after part of it may have to go higher or lower than the fore part, in order to gain some advantage; but this change may be effected, if necessary, without in any way altering the character of the line already drawn, it is only necessary to alter the height at which it should be placed, and not even that, unless required.

Sheer plan. (3). *The Sheer Plan.*

This gives the entire outline of the ship as we look at her sideways. The top line, or upper boundary, is the line of her deck or bulwark, or in short, the top of the ship, which must be laid down in order to construct the chief buttock-line in section (4); the bottom is the line of her keel, the front is the line of her stem or cut-water, and the after part is the line of her stern-post.

Stem. Begin with the stem:—Make that line follow the form of the chief buttock-line and gradually grow out of it. This ought to be so, because buttock-lines bound equal thickness of the ship, and a stem is merely a thin slice of the ship, and therefore, follows one of the buttock-lines. As a matter of beauty and of reason, therefore, it must be made a buttock-line in order that the outline may harmonise with the general form.

HOW TO DESIGN THE LINES OF A SHIP.   89

The form of the stem, therefore, will depend on the decision taken with regard to the deck-line, and the buttock-line. If the deck-line be kept well aft, so that the main buttock-line tumbles home, the stem above the water will be curved backwards, and so be in unison with the tumble home bow, and it will be the contrary if the clipper bow be adopted. Above the water, however, the mere form of the stem itself is a matter, to some extent, of taste or fancy. If the general character of the bow give good buttock-lines, it will not matter much whether the stem to which they are joined curve out or in, except that it will always be better for the stem to harmonize with the character of the bow of which it forms the conspicuous outline. Some constructors hesitate to give it a decided character, and express their imbecility by leaving it perpendicular.

Below the water, on the contrary, the form of the bow is of the greatest practical value. It has been the custom to carry the stem down to an angle with the keel, to continue the keel forward to meet the stem, and so form what is called "the fore-foot" of the ship, giving it a great gripe, or hold of the water. This gripe and fore-foot have every bad quality, being weak in structure, and making the vessel hard to steer. It is thought better to cut it all off, following the shape of the other buttock-lines; and further to carry this rounding a great way back, say to one-sixth of the whole length of the ship. Rounding off the fore-foot.

By this, not only does the Constructor diminish the fore gripe and ease the steerage, but the stem and fore-foot is kept out of harm's way; and this has been known to save repairs and contribute to the safety of the ship. When turning in narrow channels, when steering in intricate or shallow waters, or performing evolutions under difficult circumstances, the fact that there is no thin, protruding part near the bottom to touch the ground, to be broken off, or to impede or alter the movement or direction of the vessel, is often of great consequence.

By a gentle curve at the stem, therefore, the fore-keel and fore-foot are kept out of harms way, and the same may generally be done at the stern, where dead-wood can be spared. By curving the after part of the keel upwards, like the stem, both keel and rudder are often saved and the ability to turn the ship certainly facilitated; of course it is done to a much less extent at the stern, as several feet of gripe at the bow will correspond with a few inches of trim by the stern.

But, though the keel and fore-foot are thus curved, the whole central part of the keel should be kept perfectly straight, for no other purpose than to be able to support the middle of the ship on blocks in the dock. This is obviously necessary, since otherwise the keel would rest merely on points, instead of being uniformly supported; and in the dock it is rather an advantage that the two ends of the ship should not be borne by the blocks, unless they require special repairs, when they can be then propped up, as may be needful. Middle part of keel to be kept straight.

It is not necessary that the keel should be parallel to the water-line, except where there is a narrow limit to the extreme Trim.

draft of water, in which case, the keel should be parallel to the load water-line; in most other cases it should incline downwards at the stern, so as to draw more water abaft than forward.

In the case of screw steamers of great power, it is a necessity to have this draft, in order to get the screw sufficiently large and sufficiently under water for effective power; and further, it is convenient in sailing vessels, to be able to carry a large sail area on the after part of the ship, for which greater depth of keel aft than forward, affords the necessary facility. It is convenient, frequently, for the same purposes, that a vessel when light, should draw very much more water aft than forward, and that her lading should bring her down gradually to an even keel. This greater draft aft than forward, is usually called "the difference of the ship," and it is reckoned a main element in her trim.

Rake of the stern-post. Another element of the sheer plan is the rake of the stern-post, and great license is allowed here. The Constructor may have the stern-post straight up and down, so as to make the rudder pivot fairly, or he may incline the head of the rudder back behind the perpendicular at an angle, or he may incline the heel of the rudder forward of the perpendicular; indeed, he may make the line of the rudder cross the perpendicular at any angle he may choose. In the first case, he will maintain the balance of his original draught. In the second case, he will extend the deadwood and increase the lateral resistance to leewardliness. In the third case, he will decrease the lateral resistance, but increase handiness. In every intermediate degree between these two, he will gain one of these qualities at the sacrifice of the other.

Rudder. The effect of this inclination on the rudder itself must not be forgotten. The inclination of a rudder increases its power to turn the ship, but it also increases the resistance which the application of rudder offers in every degree to the progress of the ship through the water. The action of the rudder, as has been already stated, is of the nature of a hindrance to one side of the ship, so as to allow the other side to go forward at greater speed thus turning the ship; but the inclination of the rudder-post has a double effect, by which, when the rudder is held over, not only is one side of the ship hindered, but a certain quantity of the water which strikes the rudder is diverted upwards, as well as to one side. Nevertheless, a certain amount of rake should be given where very great power of rudder is needed.

Counter. Next comes the question of rake of counter, and rake of stern. It is thought best to allow the buttock-lines to decide the rake of the counter, so that when the stern is deep in the water, the counter may be a continuation of the true form of the ship and of her lines. A good counter of this sort, will help the ship instead of hindering it, when the stern happens to be buried in the waves. As to rake of stern, it seems to be a matter of fancy, especially in this country. It is better that the stern should rake outwards, rather than it should tumble home.

Sheer. The upper boundary line, which appears to finish the ship, is called "the sheer-line," and this, also, is a mere matter of taste, though some points of it have more or less reason. In looking at a ship which has little sheer, she is apt to have the appearance, contrary to the truth, of being rounded down, that is to say, as

HOW TO DESIGN THE LINES OF A SHIP. 91

if drooped at the ends. Now this is universally agreed to be so ugly, that a considerable sheer at the bow and somewhat less at the stern, are necessary to counteract it. Experience gives for a vessel of 200 feet length, a sheer or rise of about 2 feet at the bow and about 8 inches at the stern. But this is frequently exceeded.

So much for the quantity of sheer. The quality of it depends on the exact curve which may be adopted. A parabola is deemed best for the sheer curve, and to trace it, proceed as follows:— Dividing the vessel into ten equal parts, six forward and four abaft, rise forward successively 1, 4, 9, 16, 25 and 36 inches, and abaft 1-2, 2, 4 1-2, and 8 inches. This gives a total spring of 3 feet forward, and of 8 inches aft, and makes the bow 28 inches higher out of the water than the stern. This proportion will serve for vessels over 200 feet in length; but for smaller ones it would be an excess. Nevertheless, it is to be observed, that even in very small vessels, especially when low on the water, a considerable sheer forward is useful to keep them dry. *Sheer line—how drawn.*

The sheer line is important in its structure thus far, that it is usual to make the planking of the upper part of the ship, the line of the ports, and the line of the decks, follow the line of sheer, though it is sometimes convenient to deviate from this usual rule, and to make the decks follow any line that may be convenient for the internal arrangements.

For example: when a large and roomy forecastle is needed, without deforming the ship by raising the forecastle above the bulwarks, it can be obtained by running the line of the deck straight forward, on the level, and so following the level of the water-line instead of the sheer of the rail; in this way, the height of the top of the bulwark above the deck, which amidships might be 5 feet, might be 8 feet at the stem, and there is no practical inconvenience from this, which is not more than compensated for by strength and usefulness. *The sheer line of the bulwarks need not rule actual construction.*

(4.) *To Construct the Chief Vertical Longitudinal Section or Buttock-line.*

In the construction of this line, there is much room for judgment; for though it does not possess such properties of its own as the water-line and midship section, it has the power of either increasing the good qualities or aggravating the evils which the ship will derive from those two primary lines. It is only secondary in importance to these, because by its means all the possible good springing from the others may be favorably developed, marred or neutralized. It happens, also, that this has not heretofore received the attention it deserves; in many designs it is not even to be found. It is believed that its good qualities tend materially to the ease, dryness, comfort and safety of sea-going ships. Inland vessels may afford to neglect it, but a practiced eye can detect in the faults of this line almost instantaneously the bad sea-going qualities of a defective design. *Chief buttock line.*

The chief buttock-line should be placed in a vertical plane parallel to the plane of the keel and the perpendiculars or central plane of the ship, and at one-fourth of her breadth from the plane on both sides. *What it is.*

Depends on the cycloid. In ordinary ships this line will be found to be of a most variable, vague and nondescript character. The wave theory adopts for it the vertical line of a sea wave, and it is thought that its conformity to that shape has everything to do with the ease of the vessel at sea. The vertical section of the common sea wave is the common cycloid. This must be elongated for a long, low vessel, and compressed for a short one. Three points through which it must pass, have already been determined by the midship section, and by the water-line, because, as this line is distant from the centre one-fourth part of the breadth, it must cross those three lines where they cross this vertical plane. These three points are the only ones which do not admit of a free choice, and it remains a part of the skill of the Constructor to adopt such a cycloid as may consist with his general design and with the use of the ship. Each of the three midship sections given, places the bottom of the buttock-line at a different depth under water, and each of the three requires a different cycloidal line to fit it. The nature of this cycloidal line has been long known to mathematicians as the only line in which a pendulum can so swing that its vibrations, whatever their extent, shall be equal-timed. There is a remarkable analogy between the swing of a pendulum and the roll of a ship; there is an equally strong resemblance between the forces which exist in a wave and the forces which act on a pendulum; the mathematics of a wave and of a cycloidal pendulum are nearly identical.

When, therefore, it was discovered that the forces which replace the water in the run of a ship are of the same nature as the forces actuating a wind wave at sea in the vertical position, by this discovery came the key to the vertical lines of the afterbody of the ship; so the vertical lines of the fore-body were contrived, in the belief that wind waves coming into collision with a body already perfectly fitted to the form which they themselves take in undulating, unresisted free motion, would not be broken, but would have free way, and that they would glide as smoothly over the face of a solid cycloid as the layers of the same wave glide over one another.

When, by experiment, the question was referred to the waves, it was proved to be so, and a vertical cycloid thus became the buttock-line of the bow of an easy and dry ship above the water, just as it had already become the easy run of the wave of replacement in the stern of the ship.

How drawn. The chief buttock-line is, therefore, described in the following manner: The after part is formed from a semi-circle, the bottom of which, is at the intersection of the midship section with the vertical plane, and of which the uppermost point is as high out of the water as the Constructor may choose to carry the bulwark. From this describe a cycloid, and cut off as much of the cycloid as may be desirable to adapt the portion of the stern beyond the perpendicular, a point which is a matter of room and comfort merely. There is choice as to whether the bow shall much overhang the water, or rise up pretty square, or tumble home. For a vessel low in the water, the first might be adopted; but never for a vessel high out of the water.

The tumble home bow is thought to be the dryest and easiest in a sea; but there is the vertical cycloid between the two. Each proportion and kind of vessel has its corresponding cycloid.

(5.) *On the Main deck line.*

By the main deck line is meant here the outline of that deck which is meant to be kept in all circumstances well out of the water. It is this which constitutes the chief gun deck of a vessel of war—on which it is necessary, in all ordinary weather, that the ports should be open, without the sea entering. There have been men-of-war in which this deck was generally *under-water*, and after long experience, they gained the name of "coffins." Main deck line.

The choice of a deck line, has a great deal to do with the usefulness of a ship for its purpose, more even than her behavior at sea. This main or construction deck is, in small vessels, the uppermost or spar deck; but in larger vessels there is a spar deck above it; in the old three deckers, there were three decks above it, and in the "Great Eastern," there are four decks above it, and *four below*. As a general rule, also, when a vessel is deeply laden, this deck is an eighth or a tenth of the beam of the ship above the water. Importance of the main deck line.

A little consideration of the purposes of a main deck, will serve to indicate how various its shape may be. In a vessel meant to be fast, its point should be like the bow of the ship, fine and sharp, because, if a full bluff deck is put on the top of a fine fast bow, the ship is given the bad quality of pitching in a sea way—the fullness of the deck line will also take from the speed—counteracting the very quality intended to be gained by the sharp bow under-water. Proper form of it.

The argument in favor of sharpness, seems inconsistent with a roomy deck forward, which is usually obtained by a broad bell bow, flaring out wide over the water. Such a bow the old school still believe in, and modern constructors would never have succeeded in introducing the fine sharp deck, in opposition to traditional prejudice, had it not been that the full deck line was found fatal to speed.

There can be no doubt, that in fine weather, a large roomy bow on deck is convenient for doing the work of the ship comfortably and handily. It is far more convenient in the man-of-war, than in the merchant ship, since in chasing, it is desirable to work two long chase guns through the bow ports, clear of everything, and to work them well in that position. It has been pretended, that it was impossible to do this on a sharp fine deck line, but this has proved a crotchet of the past.

The simple fact is, that the roominess, dryness, and comfort of a full deck line, instead of a fine one, is mere impression or belief—nothing more. If it is imagined, that a fine bow is got by cutting off so much room from a full bow, and so diminishing the extent of available deck room for working ship—then the fine bow may be considered narrow and confined; but the practical fact is the contrary of this. The fine deck line of a modern fast ship, is *not* got by cutting anything off the length, or off the width, or off the roominess of a deck; the sharp bow is Roomy bow does not require bluff lines.

obtained by adding on a fine entrance to a bluff one, and by lengthening the deck; the full parts of the ship and of the deck remain where they were. All that is necessary, therefore, is to see that the working parts of the ship shall, in the fine bow, be kept well aft, in the broad open space of the deck, and not crammed forward into the narrow space superadded—which should be kept perfectly clear. It is a further peculiarity of the fine bow and deck line, that the foremast stands much farther aft than in the old full bow, and that there is therefore more room forward of the mast; care must, therefore, be taken to keep windlass or capstan, catheads and anchors, and all working parts of the bow, well aft—not for room merely, but also to keep heavy weights out of the extreme bow of the ship, as they are always detrimental.

There is another way of looking at this matter. It is a good plan to cover in the whole of the fine part of a deck forward, with a light forecastle, bulkheaded off, especially in iron ships. It is a great convenience, and affords good quarters for the crew; it keeps the head light and dry; while abaft the forecastle, a broad roomy deck is still to be found. There is, however, another way of giving a roomy deck on a sharp bowed vessel, and it has been tried with success in men-of-war. An extremely fine bow has been made to carry two long 8 inch guns, parallel to the keel, through two wide ports, with ample room all around, to train and work them freely. This was accomplished by shortening the deck, or stopping it very much short of the bow, carrying the bulwark round the bow, considerably behind the stem; the real deck beyond the bulwark forming part of the head, which, instead of being grated and overhanging the sea, had a solid oak deck over the greater part of it, leaving the head as convenient as before. In this way the bulwark of the deck left the real line of the ship 30 feet short of the stem, with a fine, round, roomy deck, to delight an officer of the old school, by giving him all he wanted on the inside, without impairing the form which the sea had to have on the outside.

Roomy deck may be got with a tumble home bow. There is yet another way of placing a full, round, capacious deck line on a fine, hollow, fast water-line, and yet perfectly reconciling them one to another—so as to form a handsome, symmetrical sea going vessel. This is to carry out the tumble home bow—which makes a vessel dry, easy and safe. To carry out this system, it is only necessary to take a tolerably full, easy deck line, composed of two circular, or two parabolic arcs, laying them over the water-line, and so far behind it, as to be easily reconciled with it, by means of the cycloidal buttock-line; a process which will be guided in a great measure, by the point at which the cycloidal buttock-line, already drawn, meets the level of the deck.

Arrangement and form of stern. Large, capacious, roomy sterns are part of the wave form, and though apparently unsightly, give great room with less cost and sacrifice, than any other part of the vessel. A small, handsome, light stern, may be pretty as an eye model, but it is a costly whim. There are no good qualities in a ship which are not improved, and no economy which is not enhanced by a large roomy stern and deck line. In a merchantman, it gives large

HOW TO DESIGN THE LINES OF A SHIP. 95

passenger cabins, airy, as well as roomy, and in that part of a ship which pays the owner best. In the ship of war, it gives a fine roomy poop, and plenty of space for working stern guns—which, however, should seldom be required in a Yankee ship of war. But the roominess and fullness of the stern, in the neighborhood of the deck line, is the greatest element of safety in that most perilous of positions; scudding in a heavy gale and sea—and in most cases, may be used with advantage, to embrace the stability and sea going qualities of the vessel.

The best way to turn the stern to advantage, for room and wholesomeness, is to carry the breadth on deck well aft, to taper the ship in towards the stern but little, and even if necessary, to carry the projection of the stern a good way abaft, and beyond the perpendicular, following, however, and not extending beyond the vertical buttock-line already given. Here may be taken a great deal of room from the sea. But then a question arises—shall the stern be round or square? The answer is, that its bulk is the main point; its shape is of less consequence. If, as a matter of taste, the corners are cut off, it becomes a round stern; nothing, too, is more common than to see constructors cut off the stern inside, and then stick quarter galleries on the outside, to make up for the corners cut off. When little is cut off it is usually called "an elliptical stern," although it never is an ellipse; and when much is cut off, it is called "round," though it never is circular. So far as the qualities of the ship are concerned, the precise outline of the deck astern is of little importance. *Square and round sterns compared.*

The constructor is now prepared to adopt a definite form for his deck line, which is plainly a compound affair of policy and taste. For a trial line, it is thought best to use forward two arcs of a circle, intersecting at the bow, and having their centres on a line drawn athwartship, half way between the perpendiculars; thence inclining by two parabolic arcs, gradually narrowing to the breadth of the intended stern; and for that breadth, the constructor should adopt, at the point where it passes the perpendicular, some specific proportion—6, 7 or 8 tenths of the midship breadth; finishing with whatever straight line or curve may have been determined on, as regards room at the stern. Indeed, in a vessel of no great length, and without much overhanging counter, no harm is likely to arise from carrying the full breadth of the deck amidships right aft to the stern, with merely sufficient curvature to give an agreeable line. *Summary.*

The completion of the design now requires that these four ruling lines be reconciled with one another. In this operation, what the constructor must keep mainly in view, is to extend as far as possible, through all the remaining lines of the ship, the good qualities which have been established in the ruling lines.

(6.) *To Construct the Remaining Water-lines.*

It is most desirable that the water-lines of the entrance should be as exactly as possible of the same form, on reduced breadth, as the main water-line. There will be some difficulty in doing this, especially near the keel; and the tendency of these lines will be to elongate themselves forward. This is to be avoided. The remaining water-lines of the after body are to be construct- *Secondary water-lines of fore body.* *Of after-body.*

ed on nearly an opposite principal. They are to deviate rapidly from the chief water-line of the after-body already drawn; and this they will do naturally, because the main buttock-line which rules the after-body, compels the water-lines to increase rapidly in fineness as they go down in the water, and to extend rapidly in fullness as they rise to the surface; thus giving what is believed to be the best kind of stern, namely, very fine below, and very full above.

In this respect it is a contrast to the bow, which is kept as full as may be, consistently with the chief water-line all the way down.* It is desirable to have at least three complete water-lines, in order to form a first approximation to the complete calculation of the ship.

(7.) *On the Completion of the Vertical Cross Sections or Body Plan.*

Secondary cross sections. The cross sections are all to be regarded as midship sections modified, but each of them giving to the part of the ship where it lies, qualities which either enhance the good qualities of the midship section or impair them.

A vessel with a fine, powerful midship section, may easily be impaired by weak extremities, and a weak midship section may be reinforced by good cross sections, especially in the after-body.

What the designer has to bear in mind, then, is to study how far he can enhance, support, and carry out the qualities of the main midship section in the rest of the body. In this he will be materially aided by the choice which he makes of that cross section which passes through the after perpendicular. To this frame, being absolutely out of the water, he may give any shape he pleases; and having fixed this, he will find, that with the main buttock-line it rules the entire form of the after-body, and also controls materially the surface of the water-line of the stern. It is this stern cross section which should be made very full, in order to turn the after-body to the best possible account. But this fullness must not be abrupt; otherwise, when rising and falling in the sea, the counter may at times strike the water with violence.

The circumstance, that this portion of the vessel remains so entirely subject to the will of the designer, makes it, for the inexperienced, the most difficult to decide and determine; and a greater variety of forms will be found in the region of the stern above water than in any other part of a ship.

General form of cross sections of after-body. The vertical sections of the after-body followed out in the manner indicated, will be found, as they approach the stern, to have become very fine below and very full above, and so they they should be; but in the bow, there will generally be found a similar tendency of the lines to become extremely fine below, and to grow full above, and there it is necessary to counteract this tendency instead of encouraging it, as abaft. The bow cross sections must, therefore, be made to maintain their full breadth well down toward the keel, and they must not spread out too rapidly at the surface of the water and above it.

---

* This—to make it water-borne.

HOW TO DESIGN THE LINES OF A SHIP.

The reason why the fullness should be preserved below is, that it is the business of the fine part of the bow, or cutwater, to displace or remove the water out of the way of that part of the ship which is to follow; and if the bow part be cut away too fine, this work will not be done, and the part behind will still have the work of displacement, with a bluffer entrance, and a shorter time to do it in, which is the same as to say, that it would then require unnecessary force, by causing unnecessary resistance. The main water-line having, therefore, already rendered the bow sufficiently fine for the service of dividing the water, care must be taken not to carry this fineness further than necessary, or than it is carried in the chief water-line.

Much care will be needed to prevent the cross sections of the bow from flaring out very much, to meet the line of the upper deck. To avoid this, keep that line fine and throw it as far backwards from the fore perpendicular as conveniently practicable. The cycloidal buttock-line, properly used, will help to throw the deck back, and to prevent it from spreading over the fine bow; nevertheless, it will always be difficult to reconcile the wave water-line, the full deck, and the cycloidal buttock-line; but when it is well done, it makes the most beautiful as well as the best of all sea bows. For fresh water bows, it does not matter how much the deck flares out, or how much it overhangs the water; it is in the sea that the true skill of the accomplished Architect is to be developed. *Caution.*

It is not the best voyage in fine weather, but the best behavior in bad weather, which gives reputation to the truly seaworthy ship.

LENGTH OF ENTRANCE AND RUN.

There is no principle given by the wave method of construction more important than the following: *That there is a fixed proportion between the speed for which a ship is to be designed, and the length of entrance and run which must be given to her, in order to fit her for that speed.* *Fixed proportion of length to speed.*

The importance of obtaining such definite proportions has long been felt by practical men. It was known that it was very difficult, by any amount of power, to push vessels of certain length and shape through the water at high velocity. Power and money were wasted in vain attempts to make ships of unsuitable dimensions attain high speed. Vessels were filled with boilers and machinery, designed to compel the performance of high velocity. Instances are known, where a double amount of steam boiler had been provided to compel high speed in an unsuitable vessel, and afterwards, these boilers had to be removed, the higher speed being found impossible in that kind of ship, and the highest speed of which the ship was capable, was afterwards brought out with half the power. The wave principle has produced the proportions in the table on page 99 of this chapter, the cause which fixes these proportions being obvious. The length of the fore body of a ship designed on the wave principle, must be the same as the length of the wave of the first order, which moves with that speed. The length of the after body, must be the same as the length of the front face of the wave of the second order, moving with that velocity. *Proportion of length to speed.*

98   HOW TO DESIGN THE LINES OF A SHIP.

Breadth may bear any ratio to length. The wave system destroys the old idea of any proportion of breadth to length being required for speed. An absolute length is required for the entrance and run; but these being formed in accordance with the wave principle for any given speed, the breadth may have any proportion to that, which the uses of the ship and the intentions of the Constructor require. A vessel meant to go ten knots can be efficiently propelled at that speed, if her length and form be right, whether she be 3 feet beam or 30 feet.

CAUTIONS IN DESIGNING WAVE VESSELS.

In designing wave vessels, it is necessary, therefore, to distinguish carefully *three* great elements of construction, viz: The *fore body*, the *after body* and the *middle body*. The lengths of the fore body and after body are indicated by the required speed, and if the beam is fixed, it is only by means of a due length of middle body that the required capacity, stability and such other qualities are to be given as will make the ship as a whole, suit its use. Middle body is, therefore, an element demanding the careful study of the designer on the wave system, and it will well reward his pains.

Caution against caricaturing the wave system. It only remains to notice the errors sometimes committed by the novice when designing vessels on the wave system. Finding that a hollow water-line is admissable, he rushes to the extreme and makes it too hollow, and gets increased resistance; or that a fine, long entrance is good, he makes it too long, and gets increased surface; or that a full after body is admissable, he makes it too full, and spoils the steering qualities of the vessel.

On the other hand, instead of going too far, he may stop short too soon. When the water-line near the bow is made fine, and the deck allowed to remain full, the end of the ship is overloaded, and so the value of carrying weights in the centre is sacrificed to a custom. It is most unwise not to reduce the weight and bulk carried out of the water. No error is more common than to give wave line vessels greater fineness than is required for the special case, to the sacrifice of the carrying qualities of the ship. The best way of avoiding these errors is, for the designer not to adopt the system too hurriedly, nor introduce it too largely into his first construction. Let him take the lines of a ship he has already built and only alter them in a small degree on the wave principle. He will find out thus, how far he has made an improvement, and how far he has altered the ship's practical points. Next time, he may make a further change in the same direction, thus avoiding the error of rushing to an extreme, than which, there is nothing more fatal to the success of a new method. A ship all ends, with no middle, all top, with no bottom, all dead wood with no capacity, is precisely one of those caricatures of the wave principle, of which the world has seen a great many misnamed "Clippers," in which the true purposes and uses of a ship have been lost sight of, in the attempt to gain great speed *at the expense of every quality which makes speed desirable.*

Practical use should be kept in view. To guard against such errors, let it never be forgotten, that the end of all ship building is to work out the purposes of the owner. A ship of war has to fight, and a merchantman to carry

HOW TO DESIGN THE LINES OF A SHIP.

cargo. To build a man-of-war which cannot fight her battery, is a much greater fault than to make her slow. To build a merchant vessel so as to have great speed at great cost, without the capacity necessary to repay the owner his outlay, is folly.— Freight is the owner's object, and to earn the greatest freight, is the problem submitted to the Constructor. To this object the wave principle, well understood, gives a safe and certain guide. When the speed wanted for the trade is known, the wave principle gives the length of entrance and run to obtain that speed. When the cargo to be carried is known, the Constructor can say what buoyancy he needs, and what length of middle-body will carry the bulk and weight. When the draft of water is given, he is ready to decide what form of midship section will give the stiffness and weatherliness needed. When he knows the weights to be carried and the bulk to be stowed, he must take care that he carries them where they are supported by the water, and not where, being unsupported, they weaken the ship and increase its strains. If he thus keeps the uses of his ship steadily in view, he will find the principles of the wave system a safe guide to enable him to give his design those qualities without a sacrifice of those other qualities which can alone enable a ship owner or a Government to avail themselves of his science and skill.

A Table of Proportions of Bow and Stern for Wave line Ships, designed for a given Speed.

Length of Entrance and Run.

Statute miles pr. hour.	Length of entrance. Feet.	Length of run. Feet.	Statute miles pr. hour.	Length of entrance. Feet.	Length of run. Feet.
1	.42	.3	11	50.82	36.3
2	1.68	1.2	12	60.48	43.2
3	3.78	2.7	13	70.98	50.7
4	6.72	4.8	14	82.32	58.8
5	10.50	7.5	15	94.50	67.5
6	15.12	10.8	16	107.52	76.8
7	20.58	14.7	17	121.38	86.7
8	26.88	19.2	18	136.08	97.2
9	34.02	20.5	19	151.62	108.3
10	42.00	30.0	20	168.00	120.0

The lengths increase as the square of the velocities.

Table of direct head resistance at different speeds on each square foot of a ship's way.

Power required to propel a flat fronted vessel through the water.

Speed.	Speed.	Propelling force in pounds to the square ft.	Propelling horse-power.
Knots an hour.	Feet a second.		
1	1.68889	2.85235	0.00876
2	3.37778	11.40938	0.07007
3	5.06667	25.67111	0.23649
4	6.75556	45.63754	0.56056
5	8.44444	71.30865	1.09484
6	10.13333	102.68445	1.89188
7	11.82222	139.76495	3.00424
8	13.51111	182.55014	4.48446
9	15.20000	231.04003	6.38511
10	16.88889	285.23460	8.75872
11	18.57778	345.13387	11.65786
12	20.26667	410.73780	15.13506
13	21.95556	482.04647	19.24290
14	23.64444	559.05980	24.03392
15	25.33333	641.71785	29.56068
16	27.02222	730.20056	36.34868
17	28.71111	824.32799	43.03160
18	30.40000	924.16012	51.08086
19	32.08889	1029.69691	60.07607
20	33.77778	1140.93840	70.06976

Resistance varies as the square of the velocity.

Table showing the comparative areas of water-way, resistance, and carrying power to be obtained by different dimensions under same shape.

Elements of first cost.			Working cost.	Remunerative work.
Breadth.	Draft.	Length.	Resistance or water-way.	Floating weight or tonnage.
12	6	72	60	100
18	9	108	135	337
24	12	144	240	800
30	15	180	375	1,500
36	18	216	540	5,400
42	21	252	735	8,575
48	24	288	960	12,800
54	27	324	1215	18,125
60	27	360	1350	22,500
66	27	396	1405	27,125
72	27	432	1620	32,400

CHAPTER XXII.

ON THE FIRST APPROXIMATE CALCULATION OF A DESIGN.

I. *Area of Midship Section Immersed.*

The area of the midship section furnishes the chief measure of resistance of the ship. The midship section is her fullest part, and the resistance being measured by it, it is plain that the propelling power must be duly proportioned to it, and that each foot of cross section will require a given number of pounds of force to drive it through the water at a given speed. Thus, if one foot of midship section should require 30 lbs. of force to drive it through the water at the rate of 10 miles the hour, this 30 lbs. must be supplied either by horse power, engine power, or sail power. It is plain that for each unit of section must be found a corresponding unit of propelling power. Say the area of the midship section is to contain 100 square feet, the first element then to be written down is *Area of immersed midship section.*

$$\text{Area} = 100 \text{ feet.}$$

II. *Surface of Skin Immersed.*

The skin of a ship might be thought to be so perfectly smooth and water so limpid as to slip from it, but such smoothness is imaginary, and water adheres. For a short race, boats are lubricated with grease, or polished with black lead, and ingenious mechanics have invented a plan for iron ships, of blowing a film of air between the skin and the water, to cut off the adhesion of the water. The reason why copper has been introduced is, that from a peculiar quality of that metal, sensible to the touch, friction is lessened and smoothness gained. This may be practically exemplified by rubbing one's finger over a smooth bright sheet of copper that has been steeped in salt water, it will be found to have the same slippery feeling as the side of a freshly caught salmon. *Wet skin.*

With or without lubrication, it is a fact, that water sticks to the skin of a ship, that the skin drags the water with it, that smoothness and lubrication mitigate but do not annihilate it. For wood and copper, this drag is reckoned at a loss of nearly 1 lb. of force at 10 knots per hour, and on the surface of a common iron ship, it may be as much as 2 lbs., and this loss increases with the velocity, and in high velocities is an important element not to be omitted. The surface of skin is, therefore, an element in the calculation of a ship, and adding to the work to be done, should receive separate consideration.

III. *Area of Light and Load Water-line.*

The area of the water-line is a material element in the power of a ship to carry sail, to carry top-weight, to acquire stiffness, to ride easy and *to roll gently.* It is very common to measure it by the proportion it bears to the midship section, and it is practically found to be from six to twelve times that area or more. *Area of water-line.*

Fineness of water-line. There is another proportion in which its value may be generally expressed, namely: Its proportion to a rectangle, in which form it shows how much of its area has been sacrificed to shape.

IV. *Area of the Longitudinal Section in the Water.*

Longitudinal section. This area is to weatherliness, what the area of the midship section is to resistance, only the object to be obtained is the exact opposite. The midship section should be in area small, to obviate resistance; the area of the longitudinal section should be large, in order to create resistance. Area of longitudinal section when small, indicates leewardliness; when large, weatherliness. It is quite plain that area of load water-line and area of longitudinal section have an important and close connection. If a large area of load water-line be combined with a small area of longitudinal immersed section, the vessel will have power to carry much sail, but this power will be wasted by leewardliness. On the contrary, a small area of load water-line may reduce her stability so as to prevent her from carrying as much sail as her large weatherly area would enable her to bear without unduly drifting to leeward. These three areas mentioned, are evidently bound up together in the constitution of a ship.

A given area of midship section will plainly want a large power to drive it, that large power will want a large area of load water-line to carry it, and that large power to carry sail, will want a large longitudinal area to utilize it.

To obtain out of the three the greatest aggregate of useful results in these points, without the sacrifice of higher value in other good points, is to display the skill of an adept in Naval Architecture.

V. *Volume of Under-water Body or Displacement.*

Displacement. The four former elements are areas merely, and they measure the resistance to be overcome in doing the work, and the power to be used in overcoming this resistance; but the element of displacement represents the purpose for which they all combine, namely: the movement of a large mass of matter from place to place by means of the floating power of water. The Constructor must find out what is the real mass to be moved, or what is the total volume of water which is to be displaced, in order that this body or ship may float in the place of this water. To do this, he must obtain a precise measure of the volume of the body to be immersed, and the weight of that bulk of water will exactly measure the whole weight of the ship and contents. He must, therefore, calculate the number of cubic feet which the **Light displacement.** body of the ship contains; first, up to the light water-line when she floats without load, or as it is called "with a clean swept **Load displacement.** hold," and, secondly, when she is full ladened with cargo, stores, persons and provisions for the voyage.

The calculation of displacement is a problem of geometry—the Constructor measures the bulk of the part of the ship under the light water-line, and allows one ton of weight for each 35 feet, (in fresh water, 36 ft.)—this gives the number of tons which the ship with all her parts and appendages must weigh, in order to float at the water-line required.

FIRST APPROXIMATE CALCULATION OF DESIGN. 103

Between the light water-line and the load water-line, lies another part of the ship, all of which will be immersed when she is laden to the load water-line. The bulk of this part must be measured exactly, and this bulk at the allowance of 35 feet per ton, (fresh water, 36,) will show what the additional weights are which the ship will carry at the intended load water-line. The displacement of this part of the ship, measures the load she will carry. *Difference shows the load the ship will carry.*

The calculation of the displacement is, therefore, mere mensuration, or a sort of superior kind of guaging.

Besides the absolute quantity of the ship expressed in tons or cubic feet, it is convenient to express the bulk in terms of her extreme dimensions. If the ship were a mere box, her bulk and displacement would at once be found by multiplying together her length, breadth and depth, the product of these in feet divided by 35 (or 36) would be the displacement in tons of that part of the box (or ship) immersed, and would therefore represent the weight of the whole. *Co-efficient of fineness.*

But the ship may be supposed to be a box with its corners pared off, and it will, therefore, sufficiently express the deviation of a ship from the box form if the Constructor places it so as to show how much of the box is lacking. He therefore expresses the volume of his ship by the fractions 1-2, 2-3, 3-4, or 8-10, or decimally 0.5, 0.66, 0.75 or 0.8, to show how much the bulk retained in the ship is less than it might have been if the corners had been kept on. It truly represents the sacrifice of quantity to quality in a ship of which the extreme dimensions have been determined, and the fraction is called "the co-efficient of fineness."

VI. *Volume of Shoulder or Power of the Ship.*

The shoulder has been defined to be that part of a ship which is alternately immersed or emersed as she is equally inclined from one side to the other. This inclination is here assumed as 14° on one side, and 14° on the other—which gives about one-eighth the beam of the ship as the heel of the "wedge," (one-eighth is a good proportion.) In men-of-war the shoulder is frequently called "the part between wind and water," and men-of-war in smooth water, are not *supposed* to careen more than 7 degrees, though it is very probable that in rolling, they will heel over 7 degrees more; in fact, in fast vessels 14° is by no means uncommon, when carrying a heavy press of sail by the wind. This inclination brings a depth of side under the water, equal to about one-eighth part of the ship's beam. *Shoulder.*

Now these wedges of immersion and emersion, or "the shoulders," must be accurately measured, and in a vessel of curvilinear outline, the back of the wedge will have a double curvature, requiring a little judicious geometry to guage it. *Wedges of immersion and emersion.*

The constructor having exactly measured the volume of each of these shoulders or wedges in cubic feet, converts his measurement into tons, by dividing either by 35 or 36, as the case may be, the result is one element by which to measure the power of the shoulder.

VII. *Volume of Out-of-water Body.*

Out of water body should be considerable. This is the volume of room in the ship above the water-line, or the surplus buoyancy, and is a material element in the safety and sea-worthiness of a ship. An ordinary ship with very little of her body above the water is dangerous, because, if by accident, she ships water and retains it, she will sink. When the sea runs high, the upper body is required to lift the ship over the waves, otherwise they roll over her; sometimes (as in our Monitors,) it is desirable to make vessels so low, that the sea may run freely over them; but in this case, careful provision is made that the decks are made perfectly water-tight. These vessels are however exceptional.

Effect of too deep an immersion. Merchant vessels have been lost by lading them so deeply, that in bad weather they became easily submerged. Moreover, a vessel deeply immersed, has very little lateral stability; a homogeneous body entirely immersed has none whatever. Such vessels are therefore exposed to great risk of capsizing, as well as foundering. It is therefore desirable, at the load water-line, to note what relation the bulk of the out-of-water body bears to the under-water body, as there is a certain ratio which it is desirable neither to exceed nor to fall short of. Its volume, therefore, should be indicated by a fraction, showing that it is $\frac{1}{2}$, $\frac{1}{3}$, $\frac{1}{4}$, or any other suitable part of the under-water body.

VIII. *Volume of Internal Body or Room in a Ship.*

Internal roominess. This is a very different element from the displacement, which measures the whole space which a ship occupies in the water, and the dead weight both of herself and what she carries. The volume now under consideration, represents the void left in the inside of the hull, or the empty space. The thickness of the proposed hull has, of course, everything to do with this. Iron ships have, therefore, more room in them than wooden ones, because the hull is thinner. The hull of a 1000 ton ship in iron may be reckoned throughout as about 6 inches thick all over,* but a wooden one of the same size may be taken at three times this thickness. A double bottomed iron ship takes much more room off the inside than a single bottomed one, but any well contrived iron ship is much more roomy than a wooden one.

Iron ships have the most.

Commercial profit depends on roominess. Roominess in merchant ships is a source of great profit, and for this reason it should be approximately ascertained at the outset. In all war vessels, *properly built*, there is generally room for more weights than the ship is able to carry, and in all designs of a ship, there should be an early approximation to the proportion between the displacement, (which represents the dead weight she can carry,) and the volume of internal body, (which represents the space she has for stowing those weights.)

It often happens that when this proportion is not accurately settled beforehand, a ship has a great deal of room to contain cargo, without displacement enough of under-water body to enable her to carry the dead weight of that cargo; and it may also happen that she has plenty of displacement to carry more freight, without having room to stow it. Mercantile tonnage, by

---

*This must not be confounded with the thickness of the "skin" alone, which in iron merchant vessels is seldom more than ¼ of an inch. The above includes frame, &c.

FIRST APPROXIMATE CALCULATION OF DESIGN. 105

which ships are classed, charged and chartered, is now-a-days fixed by measuring the room inside of the ship, as on this depends the "registered" tonnage placed on the register of the ship.

TONNAGE LAW OF THE U. S., APPROVED MAY 6th, 1864.

The register of every vessel shall express her length and breadth, together with her depth and the height under the third or spar deck, which shall be ascertained in the following manner:

The tonnage deck in vessels having three or more decks to the hull, shall be the second deck from below; in all other cases the upper deck of the hull is to be the tonnage deck. *Tonnage deck.*

The length from the fore part of the outer planking on the side of the stem, to the after part of the main stern-post of screw steamers, and to the after part of the rudder post of all other vessels, measured on the top of the tonnage deck, shall be accounted the vessel's length. *Vessel's length.*

The breadth of the broadest part on the outside of the vessel shall be the vessel's breadth of beam. A measure from the under side of tonnage deck plank, amidships, to the ceiling of the hold, (average thickness,) shall be accounted the depth of the hold. *Breadth of beam.* *Depth of the hold.*

If the vessel has a third deck, then the height from the top of the tonnage deck plank to the under side of the upper deck plank, shall be accounted as the height under the spar deck. *Height under the spar deck.*

All measurements to be taken in feet and fractions of feet; and all fractions of feet to be expressed in decimals. *Measurements, how taken and expressed.*

The register tonnage of a vessel shall be her entire internal cubical capacity in tons of 100 cubic feet each, to be ascertained as follows: *Register tonnage to be what, and how ascertained.*

Measure the length of the vessel in a straight line along the upper side of the tonnage deck, from the inside of the inner plank (average thickness) at the side of the stem, to the inside of the plank on the stern timbers, (average thickness,) deducting from this length what is due to the rake of the bow in the thickness of the deck, and what is due to the rake of the stern timber in the thickness of the deck, and what is due to the rake of the stern timber in one-third of the round of the beam; divide the length so taken into the number of equal parts required by the following table, according to the class in such table to which the vessel belongs:

TABLE OF CLASSES.

1st. Vessels of which the tonnage length, according to the above measurement, is fifty feet or under, into *six* equal parts. *Class 1st.*

2d. Vessels over fifty, and not exceeding one hundred feet in length, into *eight* equal parts. *Class 2d.*

3d. Vessels over one hundred, and not exceeding one hundred and fifty feet in length, into *ten* equal parts. *Class 3d.*

4th. Vessels over one hundred and fifty, and not exceeding two hundred feet in length, into *twelve* equal parts. *Class 4th.*

5th. Vessels over two hundred, and not exceeding two hundred and fifty feet in length, into *fourteen* equal parts. *Class 5th.*

106 FIRST APPROXIMATE CALCULATION OF DESIGN.

Class 6th.

6th. Vessels of which the tonnage length, according to the above measurement, is over two hundred and fifty feet long, into *sixteen* equal parts.

Transverse area of vessel, how ascertained.

Then the hold being sufficiently cleared to admit of the required depths and breadths being properly taken, find the transverse area of such vessel at each point of division of the length, as follows:

Transverse areas.

Measure the depth at each point of division from a point at a distance of one-third of the round of the beam below such deck, or, in case of a break, below a line stretched in continuation thereof, to the upper side of the floor timber, at the inside of the limber strake, after deducting the average thickness of the ceiling, which is between the bilge-planks and limber strake; then, if the depth at the midship division of the length do not exceed sixteen feet, divide each depth into four equal parts; then measure the inside horizontal breadth at each of the three points of division, and also at the upper and lower points of the depth, extending each measurement to the average thickness of that part of the ceiling which is between the points of measurement; number these breadths from above, (numbering the upper breadth one, and so on down to the lowest breadth;) multiply the second and fourth by four, and the third by two—add these products together, and to the sum add the first breadth, and the last or fifth; multiply the quantity thus obtained by one-third of the common interval between the breadths, and the product shall be deemed the transverse area; but if the midship depth exceed sixteen feet, divide each depth into six equal parts, instead of four, and measure as before directed, the horizontal breadths at the five points of division, and also at the upper and lower points of the depth; number them from above as before—multiply the second, fourth and sixth by four, and the third and fifth by two; add these products together, and to the sum add the first breadth and the last or seventh; multiply the quantities thus obtained by one-third of the common interval between the breadths, and the product shall be deemed the transverse area.

Register tonnage, how ascertained.

Having thus ascertained the transverse area at each point of division of the length of the vessel, as required above, proceed to ascertain the register tonnage of the vessel in the following manner:

Number the areas successively, one, two, three, &c., number one being at the extreme limit of the length at the bow, and the last number at the extreme limit of the length at the stern; then whether the length be divided according to the table into six or sixteen parts, as in classes one and six, or any intermediate number, as in classes two, three, four and five, multiply the second and every even numbered area, by four, and the third and every odd numbered area (except the first and last) by two; add these products together, and to the sum add the first and last, if they yield anything; multiply the quantities thus obtained by one-third of the common interval between the areas, and the product will be the cubical contents of the space under the tonnage deck; divide this product by one hundred, and the quotient being the tonnage under the tonnage deck, shall be deemed to

FIRST APPROXIMATE CALCULATION OF DESIGN. 107

be the register tonnage of the vessel, subject to the additions hereinafter mentioned.

If there be a break, a poop, or any other permanent closed in space on the upper decks, on the spar deck, available for cargo or stores, or for the berthing or accommodation of passengers or crew, the tonnage of such space shall be ascertained as follows: *When there is a break or poop on the upper or spar deck.*

Measure the internal mean length of such space in feet, and divide it into an even number of equal parts, of which the distance asunder shall be most nearly equal to those into which the length of the tonnage deck has been divided; measure at the middle of its height, the inside breadths, namely: one at each end and at each of the points of division, numbering them successively, one, two, three, &c.; then to the sum of the end breadths add four times the sum of the even numbered breadths and twice the sum of the odd numbered breadths, except the first and last, and multiply the whole sum by one-third of the common interval between the breadths; the product will give the mean horizontal area of such space, then measure the mean height between the planks of the decks, and multiply by it the mean horizontal area; divide the product by one hundred, and the quotient shall be deemed to be the tonnage of such space, and shall be added to the tonnage under the tonnage decks, ascertained as aforesaid.

If the vessel has a third deck or spar deck, the tonnage of the space between it and the tonnage deck shall be ascertained as follows: *When there is a third deck.*

Measure in feet the inside length of the space at the middle of its height from the plank at the side of the stem, to the plank on the timbers at the stern, and divide the length into the same number of equal parts into which the length of the tonnage deck is divided; measure (also at the middle of its height) the inside breadth of the space at each of the points of division, also the breadth of the stem and the breadth at the stern; number them successively, one, two, three, &c., commencing at the stem; multiply the second and all the other even numbered breadths by four, and the third, and all the other odd numbered breadths (except the first and last) by two; to the sum of these products add the first and last breadths; multiply the whole sum by one-third of the common interval between the breadths, and the result will give in superficial feet the mean horizontal area of such space; measure the mean height between the plank of the two decks and multiply by it the mean horizontal area, and the product will be the cubical contents of the space; divide this product by one hundred, and the quotient shall be deemed to be the tonnage of such space, and shall be added to the other tonnage of the vessel ascertained as aforesaid. And, if the vessel has more than three decks, the tonnage of each space between decks, above the tonnage deck, shall be severally ascertained in the manner above described, and shall be added to the tonnage of the vessel ascertained as aforesaid.

In ascertaining the tonnage of open vessels, the upper edge of the upper strake is to form the boundary line of measurement, and the depth shall be taken from an athwartship line, extending from the upper edge of said strake at each division of the length. *Tonnage of open vessels.*

Register to express the number of decks, tonnage, &c.

The register of the vessel shall express the number of decks, the tonnage under the tonnage deck, and that of the between decks, above the tonnage deck; also, that of the poop or other enclosed spaces above the deck, each separately. In every registered United States ship or vessel, the number denoting the total registered tonnage shall be deeply carved or otherwise permanently marked on her main beam, and shall be so continued; and if it at any time cease to be so continued, such vessel shall no longer be recognized as a registered U. S. vessel.

Tonnage to be marked on the main beam.

The foregoing may be simplified into a formula as follows:

$$\text{Area} = [A + 4P + 2Q] \times r\text{-}3.$$

Where $A =$ sum of the first and last ordinates,
$4P =$ sum of the even ordinates multiplied by four,
$2Q =$ sum of the remaining (or odd) ordinates multiplied by 2,
and $r\text{-} =$ the common interval between the ordinates.

A ship is said to be 1000 tons burthen, therefore, when she has 100,000 cubic feet of space. This is now the technical tonnage of the Custom House; but ship owners sometimes reckon 40 or 50 cubic feet to the ton, according to the nature of the trade that the shipper is chartering for.

IX. *Critical Points in a Ship.*

Critical points.

The Constructor has hitherto considered certain important areas and volumes of a ship, on the mutual proportions and relations of which, her qualities and powers must depend; beyond these, however, there remain certain critical points or places which are material to the whole behavior of the ship. Unless he first knows these critical points, he knows nothing about where he should place one thing or where another. Situation, is a material part of Naval Construction; masts may be placed right or wrong, machinery, boilers, coals, heavy cargo, light cargo, provisions, water, guns, anchors and cables, everything, in short, that a ship is to contain and carry, as well as every particle of weight in the hull itself, may be placed either right or wrong; and even a very small weight may be so placed as to enhance some virtue or exaggerate some defect.

X. *On the place of the Centres of Gravity of the Midship Section, and of the Vertical Sections parallel to it.*

Centre of gravity of midship section.

The position of the centre of gravity of the immersed body of a ship is a material element in her behavior and qualities at sea. The place of the centre of gravity of the midship section contributes more powerfully to determine the vertical height of that centre of gravity than any other section of the ship, and the higher that centre of gravity rises towards the surface of the water, the greater is the stability. Its depth below the surface, when the ship is light, and when she is laden, is therefore a material element in her character—and in every trial design, its place ought to be well marked.

Other cross sections.

The qualities of the ship will also be affected somewhat by the position of the centre of gravity of the vertical sections before and abaft the midship section, and to show how these rise and fall, and modify the whole, it is recommended that a line be

FIRST APPROXIMATE CALCULATION OF DESIGN. 109

drawn on the sheer plan, connecting all the centres of gravity of the vertical cross sections fore and aft. This line will be instructive, in showing the general character of the fore and after body, in improving the character of the midship section, as regards stability or in weakening it. Where the line rises, the stability is improved, where it falls, it is weakened. How they affect stability.

XI. *On the place of the centres of gravity of the Water-lines.*

The place of the centres of gravity of the water-lines form elements in the determination of the place of the centre of gravity of the ship lengthwise, and they should be carefully marked on the sheer plan and connected by a line. This line of centres of gravity of water-lines will either shift forward as the ship goes down in the water, or shift aft, or remain stationary in a vertical line; if it shift aft, it will show that, as the ship gets deeper and deeper in the water, the heavy weights ought to be stowed aft, if the contrary, then they ought to be brought forward; if neither, they ought to be equally distributed toward both ends. Thus, the line connecting these centres of gravity becomes a permanent rule for the practical stowage of the ship, exceedingly useful to the Naval Officer, and ships may thus be distinguished from each other, accordingly as these centres of gravity run forward or aft. Centre of gravity of water lines —in length.

The distance of the centre of gravity, also, of each half of the load and light water-lines, on each side of the centre of the ship, from that centre, is an element in the stability of the ship. —In breadth.

XII. *On the Place of the Centre of Gravity of the Longitudinal Vertical Section.*

As this is the section which gives weatherliness to the ship, and keeps her from drifting to leeward, the knowledge of the position of its centre of gravity, is most important to the proper placing and arrangement of all the masts, spars, rigging and sails. This centre is the balance point of the lateral pressure of the ship on the water, and the balance point of the pressure of the sails under the wind, must be so placed as exactly to correspond with it; otherwise, if the sails and masts are too far aft in relation to this point, the ship will be ardent, which carried to the extreme is a very bad quality, or if they are too far forward, she will carry lee helm, which is a worse quality. In order to secure a perfect balance, the centre of gravity of this section, and the centre of gravity of the sails, (centre of effort,) must be accurately obtained. In some vessels, these centres must be directly over each other; but in full bowed vessels, the bluffness of bow deranges the balance; and this must be corrected by carrying the centre of gravity (centre of effort) of the sails considerably forward of the centre of gravity (centre of resistance) of the longitudinal section. Centre of gravity of the longitudinal section. Balance of sail.

In one case, (a line of battle ship,) this was as much as 15 feet. In a fine, wave built ship, these centres go most accurately together.

Of course, the placing of the masts is by this means regulated and determined.

XIII. On the Place of the Centre of Gravity of the Volume of Displacement, or Centre of Buoyancy.

Centre of buoyancy.

This point must be found, because it is the centre (or balance point) of the whole upward pressure of the water on the ship; whatever may be the variety and shape of the parts of a ship, the joint power of them all, united to support a given weight, balances at this point. It has already been shown that all the weights must be so distributed as that they shall all balance each other exactly, so that their united centre of gravity shall come precisely over the centre of gravity of the displacement, otherwise the ship will be pressed down out of its intended place some way or other.

In length.

In height or depth.

The determination of the horizontal place of the centre of displacement of the whole ship, is not the only thing necessary for the balance of the ship, but the vertical distance of that centre of displacement, below the water-line, is an element of calculation in her stability.

For these two purposes, it is necessary to take, as one of the elements of a ship, first, the place of the centre of displacement, either before or abaft the middle of length of the ship; second, the distance of the centre of displacement below the water-line. The distance before or abaft is generally reckoned in feet, and the distance below the water-lines in fractions of the main breadth. The smaller this fraction is, the greater will be the stability; for the distance of this point below the water, is a measure of the tendency of the under-body to upset the ship.

XIV. On the Place of the Centre of Gravity of the Volumes of the Immersed Fore and After Bodies.

Centre of gravity of fore and after displacements affect pitching.

When a ship breasts the sea, her fore body has to be lifted by the waves, so as gradually to raise the ship out of the hollow on to the top of the wave. In lifting, the ship is said to 'scend, (from ascend,) and when she goes over the crest, and pitches down the slope on the opposite side, the fore body is raised out of the water, to plunge afterwards into the ascending slope of a second wave, until the buoyancy of the fore body and the lifting power of the second wave again raise the bow, towards the crest of the wave, and thus lead the body over a second wave. This pitching and 'scending is mainly done by the fore body; and, in order to measure and appreciate its good or bad qualities in this respect, the position of its centre of displacement should be known.

The after body also contributes its share to the movement of the ship, and its action is similar to that of the fore body in many respects, with an important difference, however, due to the ordinary forward motion of the vessel. The wave strikes the bow with a force, which the stern in a great measure escapes. But it is necessary to know the place of its centre of gravity of displacement also.

It is only necessary to remark here, that the further the centre of gravity of the fore body lies forward of that of the vessel, the greater will be the force with which the wave compels the fore body to rise, or allows it to fall; and the same, in a less degree, holds good of the after body. It is usual therefore to state, in fractions of the length of the fore body, the distance of

FIRST APPROXIMATE CALCULATION OF DESIGN. 111

its centre of displacement from that of the vessel, reckoning from that centre to the forward perpendicular, and in like manner for the after body abaft to the after perpendicular.

Thus, if the fore body were pyramidal, its centre of gravity would be one fourth, or 0.25; if ellipsoidal, it would be three-eights, or 0.375; or if wedge form, one-third, or 0.33; and if a square box, one-half, or 0.5. Position of centre of gravity of some special forms.

XV. On the Place of the Centre of Gravity of the Upper Body of a Ship, above the Water, and of its Fore and After Parts.

These are to be found and recorded in like manner with those of the under body, and for like reasons. If these are found to correspond with those of the under-water body, the ship will have this advantage, that the fore body in lifting its own top-weight, will apply its raising force exactly under the centre of weight to be lifted; and when the under-water body supports the out-of-water body, it will apply its force directly under the weight to be raised, and by this means the straining of the ship will be the least possible. Centre of gravity of upper-body.

XVI. On the Place of the Centre of Gravity of the Internal Room of the Ship, and of its Fore and After Bodies.

If it be conceived that the whole of the ship is uniformly filled with a homogeneous cargo, (say tea or sugar, coals or cotton, wool or corn,) or any other uniform weight, it is plain that no discrimination or discretion can be used in stowing the weights. The cargo being all of one sort, there is no latitude for disposing heavy or light weights, where it would be most proper for them to be carried; the condition is simply that the ship must be filled and take her chance. In such a ship everything depends upon the original design; if when she is full, she is found to be badly trimmed, perhaps down by the head, or say down by the stern—it is too late to help it. Centre of gravity of figure of the whole ship.

It is, therefore, indispensable that the Constructor should know where the centres of weight of the internal hold, stowage and bulk of such a ship lie.* He must ascertain, first, the centre of gravity of her entire internal capacity, and then of the fore and after holds; and if these fall over the corresponding centres of gravity of displacement of the ship when at her load water-line, then the cargo will exactly balance, and the trim of the ship will be perfectly maintained; if not, a difference will be manifested, which will give him the measure of how much she will be out of trim when laden. If this difference be inevitable, no remedy remains, except the empirical one of putting in some ballast; and the places of these centres of gravity will be needed for the calculation of how much ballast the ship will require, and where it should be placed. This is a sufficient reason why the centres of gravity of room for cargo should be calculated with the same accuracy as those of the centres of displacement, and in a similar manner.

XVII. On the Place of the Centres of Gravity of the Shoulders of the Ship.

When a ship heels over under the pressure of wind and sail, it is the power of the shoulder only, that enables her to stand Centre of gravity of shoulder.

* The Stevedore should know this also.

up under it, and therefore it is necessary beforehand to know, not merely the bulk of the shoulder or its quantity, but the manner of its application, more or less advantageous, to sustain the pressure it is required to withstand. A given quantity of sail may be applied high up on a mast, or lower down, and a given quantity of shoulder will require to be applied either further away from the centre of the ship, or nearer, to suffice for this effort. The centre of effort of sail and the centre of effort of shoulder have, therefore, both to be found, and both to be measured, from the central axis round which, speaking roughly, the ship has turned in heeling. This measure is generally given in feet, but the centre of the shoulder may also be reckoned in fractions of the half breadth of the ship. Thus, in a wall-sided, square ship, it would be at two-thirds of the half breadth; and in a wedge shaped ship, between one-half and two-thirds.

XVIII. On the Weight of the Hull of a Ship, and the place of its Centre of Gravity.

Weight of hull, how found roughly.
For the purposes of theoretical calculation merely, it is a good, though rough approximation, to take the whole skin of a ship, including her deck, as of a given uniform thickness and weight; and if the Constructor knows from his own experience, the total probable weight of such a hull, he will find it sufficiently near the truth, to make a first calculation in this manner. Of course it is not true, absolutely, in any case, as scarcely ever are two ships built alike in distribution of materials; but it is sufficiently near the truth to enable the dexterous ship-builder to make it absolutely true in the *ultimate practical result*, by a thoughtful distribution of all the weights of the ship, as to which he has free choice.

More accurate calculations must then be made.
All this will serve only for a first approximation; then, when all the details of the actual structure and equipment of the ship are finally settled, and everything placed, a final calculation must be made with great accuracy, showing the absolute volumes and weights of the hull, and of its different sub-divisions, and the positions in height, length and breadth of the centres of gravity, both of figure and (where practicable) of actual weight, as well as the areas and centres of the principal planes. When this is done, the Constructor will see whether his centres of gravity of displacement coincide with the position of the centre of gravity of the hull, and with those of the weights which the ship is to carry, so that the ship, as a whole, as well as each part of her, may do its work perfectly, and that her trim, on going to sea, may be found to be exactly as originally intended.

When the designer has found the place of all of these points, has measured the areas of all these surfaces, and has guaged all the volumes and capacities mentioned, he has then the elements which will enable him to judge of the qualities of his ship.

He may arrive at this judgment in two ways, either by comparing the elements thus obtained with all the similar elements of a known ship, whose good qualities he means to imitate, or he may proceed to make an absolute mathematical determination of the qualities of his ship, without reference to any other vessel.*

---
* See Chap. VIII Russell's Naval Architecture.

CHAPTER XXIII.

SHIPS FOR WAR.

A man-of-war, in general structure, may differ from a merchantman either very little or very much. A first class clipper differs very little either in size, proportion, shape or qualities from a fast sailing frigate of equal tonnage; it is mainly in the interior arrangement, fitting and equipment that they differ. Both equally require that the design of the hull shall be stable, weatherly, fast, easy and handy, and that their structure shall be stout and staunch, and that their driving power shall be such as to give them the speed required. The Constructor, then, need only consider the points in which their purposes require that they should differ. *Purpose of a man-of-war.*

As the purpose of the merchantman is to carry freight and earn profit, so the object of a man-of-war is to fight and achieve victory. Strong to destroy, comes in the place of strong to carry; but just as carrying power does the merchant no good, if it does not earn profit, so fighting power does the nation no good if it does not win victory. To win, is the work of both. The question, therefore, which underlies the whole design, construction and equipment of a man-of-war is, how to win a victory? What are the points, then, in a man-of-war which will enable her to win a victory? *To achieve victory.*

The first point is, that she has the *speed* to find her enemy. To find the enemy the ship must be faster, otherwise she may never find and, consequently, never fight her adversary.* When the enemy is found, the Commander must have the power to choose his time to fight, for choice of time and place in an action is half the victory. Above all, then, *speed* is the first condition of victory. *Conditions of victory. Choice of time and place.— Speed being essential.*

If what has just been said is true of an action between ship and ship, it is much more true of a fleet; therefore, all the ships composing that fleet should, without exception, have the same uniform highest rate of speed; otherwise, it will not be the Admiral who chooses time and place for the battle, but it will be the slow ships that decide. The presence in the fleet of a few slow vessels, may be enough to lose him the battle.

The next point essential to victory is, choice of distance.— Whether he shall engage at long or short range, often decides the fate of an action. The fast ship can, if she chooses, keep at long range out of the way of the shot of her enemy, and destroy that enemy by her guns of longer range, if she posesses them. If, on the contrary, it is the enemy which has the longer range, the fast ship can destroy the inequality of range by coming rapidly to close quarters with her adversary; so that, in either case, the slow ship is nearly in the power of the faster. The fast ship and the fast fleet then, command the victory. *Choice of distance.*

The next point after speed, is *steadiness*. A ship of war must be regarded as the platform of a floating battery; if this platform be steady, her guns may aim true, and deliver destructive fire; if not, they fire wildly and do little execution. To waste amuni- *Stability of fighting platform.*

---
* It will be remembered by the readers of Naval history, that much of Nelson's time was lost in fruitlessly endeavoring to find the French fleet.

tion is to fail. A stable platform in a ship of war, is one of the highest achievements of naval science. An unstable platform is caused, sometimes, by an undue balance of weights in the ship, which gives her spontaneous rolling motion; sometimes by a form which gives her a tendency to adapt herself to every change on the surface of the sea. But a ship may be constructed so as to give the sea the least power over her, either to make her roll or pitch.

The fine ends of the wave system accomplish the one, the round tumble home side accomplishes the other; so that as to a ship's own tendency to roll, a low meta-centre and a low centre of gravity do all that is possible to prevent *that*.

The wonderful combination of ease with stability and steadiness of platform, which distinguished the old French vessels of M. Sané, are to be attributed to the success with which he gave effect to the tumble home side, the low meta-centre and the centre of gravity.

The frigate "Constitution" was also a notable instance of this; very few ships built since, have exceeded her in these qualities.

Stability of platform, and consequent steadiness, are another condition of victory. If, in a given sea, one vessel delivers her fire with sure and steady aim, and the other fires wildly, victory cannot long remain matter of doubt.

Size an element of stability. As size of ship has everything, in a merchantman, to do with quantity of cargo, so the size of a ship has considerable to do with successful action between men-of-war of the broadside pattern. The odds are in favor of the larger ship. This is untrue in one sense and true in another. It is taken for granted that the larger the ship the heavier the battery; and it is by the weight of the broadside she can deliver in a given time, that the power of a ship of war is measured. Between two ships then of different size, it must be supposed that the weight of the broadside is proportioned to the size of the ship, or that the greater tonnage carries with it the heavier armament, and the larger ship's company to work it. This being so, the victory will be on the side of the larger ship.

Power of battery. *Power of battery* is, therefore, the next element of victory, and the best ship is that which can carry and work the most powerful battery. But now the question arises, what shall be called the most powerful battery? Shall it be the greater number of guns, the greater size of the guns, or the greater weight of broadside? It is now generally admitted that victory lies with the larger guns and the heaviest broadside, and that a number of guns of small calibre are worthless.

Height of battery. The basis of construction, then, of a man-of-war, may be said to be the weight of armament she is to carry, and the speed at which she is to carry it, just as in the merchantman, it is the weight of cargo she has to receive, and the speed at which she has to deliver it; the object of both being to carry a given weight to a given place. But the difficulty with a man-of-war is, that she has to carry her weights in the wrong place; the merchantman carries her weight in her hold; the man-of-war has to carry hers on her decks. The deck of the man-of-war is her battery plat-

SHIPS FOR WAR.

form, that her guns may be carried well out of the water, which is an obvious condition (except in certain peculiarly built vessels)* of fighting them successfully. Men-of-war with a low gun deck, must shut in their ports in a heavy sea, and that deck is useless. The loss of the lower gun deck in an old fashioned line-of-battle ship was only a partial disarmament, but in the modern ship it is total defeat.

The armament to be carried, and the height at which it is to be carried, must first be settled then;—4 and 5 feet out of water was the old style; 6 to 7 is the height of the ports of the ships of the French Navy; 8 to 9 of the ships of the British Navy; 11 feet was the height of the U. S. Steam Frigate Merrimac's midship port.

That, therefore, may be called the ruling condition of the modern fleets of the broadside type.

The weight of the heaviest broadside ship of to-day, may be taken at 50 guns, of, say 12 tons each, or 600 tons of weight carried in a single tier at 9 ft., or in a double tier at 13 feet. This is the problem that English and French Architects have had to solve. How to carry such a tremendous weight steadily at the given height? How *they* have solved it—the splendid ships in their respective Navies are witnesses. *Weight of modern battery.*

Next comes a new condition of Naval Construction arising out the modern invention of iron armor. You can destroy the enemy if your ship has the speed to catch him, and battery enough to smash and sink him; but an important question remains, your powers of endurance and his. "He who stays wins." Therefore power to endure the enemy's fire, is next in value to the power with which you deliver your own broadside. The assistance of iron is therefore sought. With iron armor, the ship may endure the enemy's battering, and in this case, power of endurance is ultimate and sure victory. *Power of endurance.*

The endurance of iron armor is found to consist in two qualities, and only two: weight and toughness; without weight in the armor it is impossible to stop the moving weight in the shot, and without toughness, it is impossible to hold on against the shot for a sufficiently long time to arrest its speed. The weight of the armor struck, diminishes the speed of the motion communicated to it, and the toughness of the armor serves to spread the motion around the point struck, and to extend this motion forward along with the ball, so as to retain hold of it with most force through the longest time. This is the whole virtue of armor. Light armor is of no use, because, in proportion to its lightness, it receives more motion; rigid or hard armor is of no use, because it cannot spread the impact of the shot and keep hold of it long enought to arrest it. Hence all sorts of shapes of armor—all attempts to use thin armor—all attempts to use hard steel or tough plastic iron to arrest the shot, have failed; the part of the armor struck by a round shot has to be at least as heavy as the shot itself to keep it out, and at least so tough as to spread the blow over an area two or three times its own diameter, and be able at the same time to yield and bear without fracture, an indent nearly the thickness of the plate itself: these qualities attained, the armor is shot proof. The question of ar- *Endurance of iron armor against round shot.*

---
* The "Monitors" for instance.

116   SHIPS FOR WAR.

mor is, therefore, shortly this: It should be two-thirds, at least, the diameter of the round shot fired at it, and should be of that tough and plastic material that Sheffield and Pittsburgh have produced so successfully.

Armor plating against square-headed rifle shot, like the "Whitworth."
But when cylindrical bolts can be fired with the same velocity as round shot, still heavier armor and new conditions will be required to resist it; but there will still remain this question: Whether the punched holes of the rifled shot will do greater harm than the battering and punching power of the heavy round shot like the 15 and 20 inch? The problem of endurance may be considered as solved, when ships can be coated with armor which hardened spherical shot, of the above calibre fired with an initial velocity of 1500 feet per second, cannot pierce.

Against speed of shot and weight.
Power of endurance, therefore, is bound up in these three things: weight and quality of armor, and weight and speed of shot, with which also go weight and size of gun. For the purposes of Naval Construction it may be given as a general rule, that the thickness of the armor should be at least two-thirds the diameter of the spherical shot, and the gun about a hundred times the weight of the shot.

The 8 inch shot will have to be stopped with $5\frac{1}{2}$ inch armor, the 9 inch shot with 6 inch armor, the 12 inch shot with 8 inch armor, the 15 inch shot with 10 inch armor, and the 20 inch shot with 15 inch armor. These are the conditions of endurance to be met at present.

But armor to a ship, like armor to a soldier, is plainly an encumbrance and embarrassment, as well as a defence. It is a great weight to carry; it is top weight, and therefore hard to carry; it is winged weight, and therefore slow to move, but hard to stop when moving. Armor, then, adds a new difficulty to construction, of the same nature as a heavy battery of guns, with this addition, that it is much greater in quantity.

Weight of armor.
The armament of the side of one of the English iron clads for a single gun only, when 5 inches thick, weighs 30 tons; and if a two decker, 20 tons for each gun. It is plain, therefore, that a ship which was able to carry 5 ton guns on its deck may be quite unable to carry those guns with the addition of 20 or 30 tons of armor for each gun, and assuming that a 15 ton gun is the future armament of a broadside ship, she will have to carry with each of these guns 60 tons of armor, if a single decker, or 40 tons of armor if a double decker!

Necessities of construction arising out of armor.
These considerations enable one to understand and measure the work to be done by an iron plated armor ship. If her guns be light and numerous, and her armor thin, she may be able to carry it over her whole length, with only so much additional breadth as suffices to carry the weight of armor and armament—to carry it in respect to buoyancy and in respect to stability; but when the weight of the guns and the thickness of the armor are increased so much that the dimensions to which the Constructor may be limited are inadequate to carry them, he has to begin afresh, and seek new conditions of structure.

Partial battery system.
It is thus that the partial battery system has grown out of extreme weight of battery and of armor; the size of a ship is

SHIPS FOR WAR. 117

limited by the narrowness and shallowness of the channels it has to navigate, and by sea-going qualities. Docks and harbors sometimes limit these dimensions, but it is less costly to alter docks or dredge harbors than to have a fleet over-matched and defeated; and it is agreed that the whole question is one of gaining victory.

If, therefore, the Constructor has a given length, width and depth, it is clear that these dimensions limit the power of the ship to the weight of guns and armor she can carry; and it is no longer a question, as in the old wooden ships, of carrying her battery or gun platform along her broadsides from stem to stern; so many guns and so much armor as she can securely carry, she may take, and to that quantity she is limited by the conditions of her existence. These conditions have given rise to the modern system of partial batteries, of which the "Ironsides" of our service, the "Warrior" and "Achilles" of the English Navy, and the "Magenta" and "Solferino" of the French Navy, are notable instances.

The "Warrior's" gun platform occupies only 220 feet of her length; the "Solferino's" double tier gun battery only 150 feet of her length; the "Bellerophon's" only 150 feet on a single deck.

Necessity, therefore, and not choice, has been the origin of the partial battery system; and the sailors of past navies, who regret the continuous batteries of the old wooden fleets, should remember that it is simply impossible to carry a greater weight higher out of the water, with stability and sea-worthiness, than the laws of nature will admit. Grows out of the necessities of iron armor.

There are two important considerations belonging to the partial battery system—one, is the great stability of platform and admirable sea qualities which arise from the concentration of great weights on the central body of the ship, instead of carrying them out to the ends. Practical properties of the partial battery system.

To a modern fast screw steamer fine ends are indispensable; to cover these fine ends with heavy armor and broad gun platforms, is to produce in every way a bad sea-going ship; no more armor must extend towards the ends than is indispensable for the ship's endurance; and, therefore, it is enough that it be carried up to the first deck out of water, and as far forward as the ship needs protection; this done, the armament and armor should be concentrated in the centre, where the middle body has power to support them, and where action of the sea on the ends of the ship will not much disturb their stability. The battery thus concentrated in the middle, the extremities of the vessel may serve for all that accommodation for officers and crew, which can only be well given where there is plenty of room, light and air; and it is only this system which can thus combine in the same ship an impregnable fortress in the centre, and *a roomy, well ventilated home* in the two ends. The partial battery system, therefore, is the best solution of the problem of heavily armed, iron clad, distant cruizing, speedy and sea-worthy ships.

The form, therefore, which the problem of modern naval construction takes, is this: Such a length of ship as will enable her to attain the speed necessary to catch her enemy, choose her time and place of action, and fix her own distance for engage- Conditions of construction on that system.

ment. The dimension of her breadth will next be fixed by the height at which her gun platform must be carried above water, while the number of guns which that platform must contain, should be limited entirely by the quantity she can *steadily* sustain. The partial battery thus becomes a fortress, within which must be included whatever is most vital and valuable—guns, ammunition, engines and boilers.

Turret system. One of the forms of partial battery, is the circular form or revolving turret, which deserves separate notice, since a kind of rivalry has arisen between it and the fixed partial battery; but in reality, there is no antagonism, both being parts of the same system, capable of being used together or separately, under peculiar circumstances, to which either is the better fitted. The case to which the revolving turret is peculiarly suited, is this: A ship has not always the power of running close to her enemy, and sustaining his fire with closed ports, until she is fairly alongside and ready to deliver a broadside. The case is frequently that of her having to chase an enemy, out-manœuvre her, or pass a battery, through a sinuous course or tortuous channel. In such a case, there occur many positions of the ships where broadside guns are of no avail—not having the lateral training sufficient for the purpose.

Its peculiar advantages. The revolving turret has therefore the following advantages: It supplies a convenient and easily handled mounting for a very large gun; it has machinery which enables that gun to be worked with a very small crew, trained with the greatest ease, and aimed with the greatest exactness round any number of degrees of a circle, so that its aim may be, at all times, independent of the course of the vessel; and it secures these advantages with a very small opening of port, and therefore with comparative safety to the gun's crew, by carrying round with the gun, on the same revolving platform, a complete shield of armor. It is, in short, a revolving round tower, containing a couple of guns, or a single gun, on parallel fixed platforms on the inside; these guns having no lateral train of themselves, but merely elevation and depression. The training is done by machinery, with steam from the boilers, which carries this turret round, a centre-balancing pivot or spindle directly over the keel of the ship.

The word of command diverts the turret to the right or left, slow, quick or stop—while the Captain of the gun stands, lock-string in hand, with his eye on the sight, ready to fire at the instant his gun bears upon the enemy. A trial of this arrangement through a great war, has convinced most officers that this is the perfection and luxury of gunnery.

Comparative advantages of turret and broadside vessels. But the turret system is valuable only in special cases. It enables a vessel to carry a greater weight of iron, to protect her guns and hull—on the other hand, the guns are few in number, and their fire extremely slow. The turret ship cannot afford to throw away a shot, and must come to close quarters to fight, and unless she possesses *great speed* this cannot be done. For the defence of a coast line, and for an action in which ship is pitted against ship, actual war has proved the turret system to be the better of the two. For the reduction of forts and batteries, and for distant and lengthened cruizes, the broadside system has the advantage.

The operations against Charleston, and the reduction of Fort Fisher, during our late war, abundantly proved the merits of the two systems, as illustrated in the case of the "Monitors" and "New Ironsides." The views here expressed are not therefore exclusively partisan to either system, and the following is suggested as a plan which the rivalry between the two parties has prevented both from adopting.

When the number of guns is small, say four, and the ship is of sufficient size, conveniently to carry them, let her carry four guns in two turrets, and use both systems together. Place two turrets, one in the after body and one in the fore body, afore and abaft the engine room; then enclose the whole space on deck between the two, so as to form a broadside battery, the bow turret might sweep 270° of the horizon, the stern turret the same, and in addition to the turrets in the ship, with 60 feet of engine room, have a broadside of four guns, or eight in all on each broadside; the whole length of battery not to exceed 120 feet of the centre of the ship. *[Value of using both systems.]*

The protection of boilers, magazine and machinery, and the deck immediately under the feet of the gun's crew, is by far the most important feature of protection. The protection of the gun's crew or of the battery, is not so important for the following reasons: Naval action with long range guns, will commence as soon as the ships are within range of each other, and will probably conclude before coming to close quarters. It will, therefore, be the destruction of the ship, rather than the slaughter of the crew, which will decide the battle. The ship's hull, as a whole, will be the target aimed at; therefore, the men will be quite ready to fight their guns as of old, without personal protection, provided the deck on which they stand is made safe.

For a multitude of purposes, fleet cruizers, with partial protection and great speed, would be most useful. They need never be reckoned as ships of the line of battle, nor take higher rank than fast cruizers. They could always choose *when to* fight and *when* to avoid action. They would never be expected to lay alongside of shot proof batteries, as they could render more important service without the chance of entailing national discredit.

Much has been said of protecting the water-line of a ship, as if the water-line was something tangible and defined. The water-line in a sea-way, is just what the fancy of the sea and stress of weather choose to make it; an abstraction, not a reality. The whole thin part of the bow and stern are liable to be out of water. Protection of a water-line is, therefore, a fictitious element of safety against the gunner, who watches his time to deliver his shell near the bow or stern of his enemy's ship, the moment the wave leaves the stern or bow bare. The only protection thin, fine ends can receive, is the sub-division horizontally, vertically and longitudinally by bulkheads—forming water tight compartments. *[Safety of unprotected parts.]*

SHIPS FOR WAR.

Water-tight compartments.

Having decided upon the quantity of armor to be carried, this sub-division into water tight compartments, becomes the great element of safety and endurance. It gives the means of sustaining damage to the hull for the longest time with the least danger.

Absolutely essential to a man-of-war.

To the utmost then, even at the loss of some convenience, the interior of the ship should have transverse bulkheads, longitudinal bulkheads and iron decks, *all water tight;* air tight even, if possible. The contents of all spaces should, to a great extent, be carried as it were in tanks. All openings for ordinary accommodation, should have water tight iron hatches, covers and doors; and closing all these openings, would be the first preliminary to action. Powerful steam pumps should also be furnished as a necessity for casualty.

What will be likely to happen if men-of-war are not provided with the water-tight compartments.

A vessel without such minute division, is not only liable to be sunk, but, long before that, is liable to overturn from diminished stability; but an iron armored ship with these provisions for action, ably carried out, may be regarded during a protracted action as equally proof against artillery, water and fire.

ENGLISH IRON CLAD FLEET.

		Class I.	Class II.	Class III.			Class IV.			Gunboats.			
		Ship of the line.	Frigate of "Warrior" Class.	Corvette of "Bellerophon" class.	Corvette with two turrets.	Corvette.	Corvette with unprotected stern.	Iron corvette without protection.	Partial battery corvette.	Class I.	Class II.	Class III.	Wave form turret vessel.
Length on L. W. line, in feet,		402	380	300	285	300	254	260	276	158	140	100	175
Breadth extreme,	in feet	68	58	56	52	48	40	40	40	25.5	23	22	25
Depth at side,	"	50	56	58	23.5	33.5	18	18	17	14.5	9.5	8	12.5
Mean draft of water,	"	28	26	24	24	20	12	12	12.5	10.5	6.75	4.5	8
Tonnage (builder's,)		8834.71-94	6212.42-94		4246	3557	2201.8-94	1906.36-94	1906.36-94	2076.56-94	360	217	553
Height of lower port-sill above L. W. line,	in feet	9	10	9.5	17	8	10*	10*	6.5	8.5*	8.5*	7.5*	7
Area of immersed mid-ship section,	in sq. feet	1617	1000	1075	1088.66	829.11	280	351.20	458.18	208	114	85	140
" of L. W. line,	"	21,360	17,680	15,110	11,963.54	11,121.65	7160	8362.96	8968.84	2553	2560	1794	2625
Tons per inch, immersion,		51	40	32	28.48	26	17	15	20	6	6	4.5	3.
Displacement,	in tons	12,728	7256	7000	5664	4711	1950	1500	2218	504	340	211	450
Thickness of armor plates, exclusive of backing.	in inches	5	6	6	4.5	4.5	4.5	...	4.5	4.5	4.5	4.5	...
Weight of hull of vessel,	in tons	4400	2500	1500	2800	1280	724	724	800	170	135	82	200
" total armor, inclusive of bulkheads,	"	2560	3207	2900	810	1150	300	...	650	100	50	47	64
" engines, boilers and water,	"	2080	1000	1000	600	890	355	355	355	75	45	21	75
" guns, ammunition, &c.,	"	720	400	200	280	256	125	125	135	50	25	12.5	12.5
" equipment, stores and fuel,	"	2300	1100	1250	1023	660	432	462	364	97	30	36	80
Number of men,		986	600	600	300	400	400	400	400	...	...	...	...
Nominal horse power of engines,		2000	1250	1000	800	1000	500	500	500	100	60	30	100
Depth of centre of gravity of all the weights below L. W. line,	ft.	3.23	−1.0	...	...	2.63	1	4	0	0	0	...	...
" " " of displacement	"	12	7.42	10.5	9	8.94	5.16	5.5	5.5	4.1	2.5	2	3.5
Height of meta-centre above centre of displacement,	"	14.25	16.72	13.61	...	11.15	13.55	13	12	6.43	9	9	9
" " " centre of gravity of weight,	"	5.49	8.30	...	...	4.81	3.49	12.5	6.5	2.33	...	...	...
Number of guns, protected,		76	26	20	4	20	...	...	13	...	2 partial.	1 partial.	1 turret.
Total number of guns carried,		90	40	20	4	20	10	10	13	4	2	1	1
Speed in knots, over measured mile,		14	14	13	12.5	13	12 to 13	16 to 17	13	9	9	6 to 7	12
Depth of armor below L. W. line,	in feet	5	5	5	5	4	4	...	5	3.5	2.5	1.5	...

CLASS I.—A ship of the line, iron clad.
CLASS II.—A frigate of "Warrior" class with lighter draft.
CLASS III.—The third column in this class, is a vessel having all her guns, magazine and engine room protected. She carries 18 broadside guns on her lower deck, the two fore and after-most ones being able to fire in a line parallel to her keel; she is, further, able to carry four guns more, two forward and two aft, in shot proof batteries on the upper deck.
CLASS IV.—The first vessel in this class, has only her two magazines, engine and boilers protected; she is able to carry a large quantity of canvas. The second vessel, or the "Alabama" class, is a fast clipper ship without any protection. She carries a very large quantity of canvas. Her ten pivot guns, when not in action, are housed in pairs, five and aft along centre of the deck; four on each side are able to fire in a line nearly parallel to the keel. The hull being comparatively low, she is a difficult target to hit.

* In Class IV. and all the classes of gunboats, the height of the lower port-sill above the water-line has been given as height of centre of muzzle of gun above the water-line, these vessels having no ports.

SHIPS FOR WAR.

IRON CLAD FLEET OF THE UNITED STATES NAVY.

Class.	No. built.	Length on deck.	Beam of iron hull.	Beam, extreme.	Draft of water.	Height out of water.	Tonnage.	No. of Turrets.	Inside diameter of Turret.	Height of Turret.	No. of guns in each Turret.	No. of Cylinders.	Length of stroke.	Diameter of Cylinders.	Indicated Horse power.	*Intended* Speed and Remarks.
		feet.	ft. in.	ft. in.	ft. in.	in.			feet.	feet.			ft. in.	in.		
Puritan,	1	351	41.8	50.	21.0	18	3265	2	26	9	2	2	3.0	100.	4560	15 knots, (never tried.)
Dictator,	1	320	41.8	50.	21.0	18	3000	1	26	9	2	2	4.0	100.	4500	15 knots, (made 12.)
Passaic,	8	200	37.8	45.	9.6	12	840	1	21	9	2	2	1.10	40.	400	7 knots.
Tippecanoe,	9	224	37.6	43.	11.6	15	1034	1	…	…	…	2	2.0	48.	1000	9¼ knots.
*Shawnee, (2 screws)	20	225	33.0	45.	7.0	12	614	1	21	9	2	4	2.6	22	600	9 knots.
Original Monitor,	1	173	36.2	41.6	10.0	18	…	1	20	9	2	2	1.10	40	400	6 knots.
Shakamaxon, (2 screws)	4	345	wood.	56.8	…	…	3130	2	…	…	2	…	…	…	…	15 knots, (never tried.)
Monadnock, (2 screws)	4	270	wood.	52.10	13.0	15	1560	2	…	…	2	…	…	…	…	10 knots.
Onondaga, (2 screws)	1	226	47.0	51.2	11.6	12	1250	2	…	…	2	…	…	…	…	9½ knots, (made 6.)

*This class, known as the "Light drafts," proved to be complete failures, and being built upon, were turned into Torpedo vessels.

ARMAMENT OF SHIPS IN THE UNITED STATES NAVY.

First Rate.	Third Rate.	Fourth Rate. (b)	Mississippi Gunboat.
1 150 pounder rifle, } in pivot. 1 XI inch 42 IX in. smooth bore } in broadside. 4 100 p'dr rifles. 4 howitzers.	2 100 p'dr rifles, } in pivot. 4 IX inch smooth, } broadside. 2 24 p'dr howitzers, 2 20 p'dr rifles.	1 150 p'dr rifle, } pivot. 1 30 " " 2 IX inch, } broadside. 4 howitzers,	3 IX inch. 4 VIII inch. 2 100 p'dr rifles. 1 50 p'dr " 1 30 " "
Second Rate.	**Fourth Rate. (a)**	**Monitors.**	**Powers of Ordnance.**

2 100 p'drs rifles. } in pivot,
20 IX inch, smooth, } in broadside.
2 60 p'dr rifles,
2 howitzers.
	1 XI inch smooth, (pivot.)
	1 20 p'dr rifle "
	4 24 p'dr howitzers, (broadside.)
		"Monadnock," 4 XV inch.
		"Montauk," { 1 XV inch.
		{ 1 150 p'dr rifle.
			In shot. In shells.
			               lbs. lbs.
			1st rate, 2606 2123
			2d " 1220 990
			3d " 424 343
			         { 210 183
			4th " { 294 255

CHAPTER XXIV.

THE DRAWINGS.

It is absolutely necessary in order to scientifically construct a vessel, that the designer, whether owner or builder, should thoroughly understand ship drawing. *Student of Naval Architecture should understand ship drawing.*

The first process towards building a properly constructed vessel, is to make accurate drawings of her on paper, upon a reduced scale. From these drawings, other drawings are made or "laid off," upon the mould loft floor. From these last mentioned drawings in chalk, moulds of thin deal are made, and by the help of these moulds, the timbers composing the "frame" of the vessel are cut out, and the frame put together. *First step towards the construction of a vessel.*

The drawings from which a ship is constructed are three in number: *The plans of a ship drawing.*

1st. The *sheer plan*, containing a series of longitudinal vertical sections.

2d. The *half-breadth plan*, containing a series of longitudinal transverse sections.

3d. The *body plan*, containing a series of transverse vertical sections.

The student should begin by copying a set of drawings, which is quite easy. Like other arts it requires a little practice, but after a few attempts he will become dexterous in the use of the irregular curves and the drawing pen. There are three methods of copying: *The student to commence by making copies of ship drawings.*

First. By a tracing of the original on prepared transparent paper or muslin, placed over it. *Modes of copying.*

Second. By placing the original plan over a sheet of paper, and pricking the principal points through with a fine needle so as to mark the lower sheet; guided by these points, the draughtsman can fill in the detail; a little practice, only, being necessary and an expert draughtsman requiring but a few points. Care must be taken to hold the needle upright.

The third plan is, to measure the principal points with a pair of dividers, or a scale, and to transfer these points to the copy, this is the best method for the student. After copying one or two drawings in this manner, he will become acquainted with the connection between the different lines in the plans. Some draughtsmen stretch their paper on the drawing board by the sponging process which presents an excellent surface for drawing on; but when cut from the board, the paper *invariably contracts*, and therefore the drawing will be more or less inaccurate according to the extent of this contraction. *Usual mode of copying.*

The best method is to hold the paper to the board with "thumb tacks." Suppose then, that the student is to copy a ship drawing. It is only necessary to describe the order in which the lines should be copied, the connection between them, and the method of ending them. *Mode of preparing paper.*

As a general rule, the lines representing the actual parts of the vessel are drawn in black; the water-lines or horizontal sec- *Colors used in ship drawing.*

tions below the load water-line is green; those lines which are not part of the vessel, but are of use in making the drawing, in ticked black ink; and the inboard work or profile, in red ink.

Light and shade. — The light is supposed to come from the right hand upper corner of the paper, and consequently the upper and right hand sides of a solid are represented by fine or thin lines; those on the lower or left hand, by thicker lines; thus, the lower side of the keel is thick, the fore side of the masts are thin, &c. The drawing should be completed *in pencil* before ink is used.

First line to be drawn. — The middle line of the half-breadth plan is first drawn, and from this line as a base, all breadths are measured—perpendicular to it, the foremost perpendicular, and then, at their proper *Vertical sections.* — distances, the lines corresponding to the other vertical sections are to be drawn.

Load water line. — The load water-line of the sheer plan, is drawn parallel to the middle line of the half-breadth plan, and consequently the sections are vertical to it.*

Load water line usually the base in yachts, but rabbet of keel in larger vessels. — From the load water-line all heights and depths are measured, it is therefore the base of the sheer and body plans. The rabbet line and lower side of the keel in the sheer plan are next drawn, the intermediate water-lines are sometimes drawn parallel to the load water-line and sometimes at equal distances between it and the rabbet line of the keel; in the former case there is less trouble in transferring the heights to the body plan, in the latter they are better adapted for making the calculations.

Sheer lines. — The several sheer lines must next be set off, the heights taken from the water-line on each section, and a penning batten made to pass through the points, and a line drawn along the batten.

Stem, rake of stern and counter. — The stem, rabbet of stem, the stern-post and its rabbet are next in order. The counter may be copied by drawing in the original a continuation of the rake of the counter through the water-line to some other line below it; this line transferred to the copy will give the rake of the counter. The detail of this part is then easily filled in.

Half-breadths. — The different half-breadths of the water-lines and of the sheer lines must next be transferred from the original, and a thin batten "penned" to pass through the points in each section. Some practice and considerable patience, is necessary in using the penning battens and weights on it; if the batten is too pliant, the line may not be a fair curve, and if too stiff it is difficult to confine it in its proper position.

Endings of the sheer and water-lines. — The endings of the sheer lines and water-lines in the half-breadth plan are obtained by squaring down from the sheer plan the intersection of each line respectively with the fore edge of the rabbet of the stem, or the after edge of the rabbet of the stern-post, as the case may be, to the middle line of the half-breadth plan, and from these spots set off from and perpendicular to the middle line half the *siding* of the stem or stern-post at the respective heights, these latter spots will be the endings required.

---

* This, however, is not always the case, for it is usual to make the upper edge of the rabbet of the keel the base of the sheer and body plans, but where there is much difference in the draft of water forward and aft, the load water-line is preferable.

THE DRAWINGS. 125

The middle line of the body plan is drawn square to the base, and the *half siding* of the stem and stern-post on each side of it. It is usual to make the base of the sheer and body plans in one line, and therefore, when the load water-line is the base, a continuation of it from the sheer plan will be both the base and the load water-line of the body plan also. When the other water-lines are parallel to the base they may also be continued, but when not parallel, the distance of each water-line from the base must be transferred from the sheer to the body plan. Lines drawn square to the middle line at these heights will represent the vertical height of each water-line at each section in the body plan, and on these lines the respective half-breadths, taken from the half-breadth plan, must be set off from the middle line, the several sheer lines are transferred in a similar manner. A curve passing through the points thus found will give the shape of that section in the body plan. The better way, is, to take off the heights and breadths of each section separately, commencing with the midship section which is often drawn on both sides; the section of the fore-body being drawn on the right, and those of the after-body on the left hand side of the middle line. The sections of the body plan will end at the half siding of the keel, stem or stern-post as to breadth, and at the lower edge of the rabbet of the keel as to depth in each section.

Middle line of body plan.

Water lines.

Fore body.
After body.

When the intermediate water-lines are parallel to the base, the breadths have merely to be set off on them. In drawing these sections of the body plan, the irregular curves must be used, and each line drawn in small pieces. After a little practice this is very easy, although, at first, some difficulty may be experienced in forming fair and correct curves. By tracing about a dozen body plans the student will become accustomed to the use of the curves. When the body plan is so far completed it may be necessary to *"fair the body"* by running *diagonal* and *buttock* lines. In transferring the former from the body to the half-breadth plan, the distance of each section is taken on the line of the diagonal from the middle line of the body plan and applied to the corresponding section of the half-breadth plan. A batten passed through these points will detect any unfairness in the line, which must be corrected. A *diagonal* line ends in the half-breadth plan at the height of its intersection with the half siding of the stem or stern-post in the body plan, transferred to the rabbet in the sheer plan, and squared down to the half-breadth; on this, the diagonal distance of the middle line to the half siding line of the body plan is set off, which gives the ending required.

Fairing the body.

Diagonal lines.

A *buttock* line in the body and half-breadth plans is drawn parallel to the middle line, the distances of its intersection with each section from the base, are transferred from the body to the sheer plan, and a batten passing through these spots will detect any unfairness, or, as a further proof the intersection of each water-line with the buttock line in the half-breadth plan may be squared up to the respective water-lines in the sheer plan; if the buttock line of the sheer plan does not agree with these last found points some alteration must be made until the body is fair, which is the case when all the intersecting points exactly coincide, and the diagonal, buttock and water-lines of each plan

Buttock and bow lines.

are fair lines. The forward portion of the *buttock* line is called the *bow* line.

Construction drawing. In the foregoing description, the body and half-breadth plans are drawn to the *outside of the plank*; in the ship-builder's **Ship-builders working drawing.** working drawings *the out side of the timbers* only, is shown. In this case the sections of the body plan and the water-lines of the half-breadth plan are ended by describing an arc of a circle with the radius of the thickness of the plank from the ending before found as a centre; the lines will end at the back of this arc.

Best mode of copying a drawing. The best rule in copying a drawing is to take as few points as possible from the original, and to find the points in one plan *from those in another;* by this means any error is much less likely to produce an unfair or impracticable drawing.

Example of construction on the tentative system. Having now given some general idea of the method of copying a ship drawing, the student may proceed to the construction of a yacht on the simplest or *tentative system,* in which the amateur Constructor takes an existing vessel or vessels of known good qualities, most nearly resembling that which he proposes to construct, as his guide, and proceeds to adopt that which is good in the vessel he establishes as his model, and to improve upon that which is bad.

·CHAPTER XXV.

CONSTRUCTION.

Having decided upon the following principal points, namely: *Preliminary.*
1. Displacement.
2. Length.
3. Breadth.
4. Depth.
5. Drag.
6. Rake of the stem and stern-post.
7. Height of the deck above the load water-line, amount of sheer, and height of the bulwarks.
8. Position of the midship section.
9. Form and area of the midship section.
10. Load water-line, its ratio to displacement, midship section and circumscribing parallelogram.
11. Centre of gravity of Displacement.
12. Inclined water-line.
13. Stability.
14. Size of rudder.
15. Area and centre of effort of sails, and position of latter in regard to centre of lateral resistance.
16. Angle of sail.
17. Size of masts and spars, and rake of masts.

The next thing to be considered, is, by what system the vessel should be constructed:

The common method of designing a vessel, is that which discards all mechanical rules for forming the various lines, and relies entirely on a consideration of those forms, which experience has taught are best adapted to the particular object in view. *Ordinary system.*

To enable a Constructor to design a vessel by this method (by far the most common one in this country) it is essentially necessary that he be provided with drawings and calculations of vessels of similar size already built; from these he can adapt and modify such parts as he considers are applicable to his purpose. When the rough drawing is made, the displacement, areas of midship section and load water-line, and the position of the centre of gravity of displacement must be calculated; should the results differ materially from those of the precedents, the Constructor must consider the probable effect of such difference upon his intended vessel, and then make such alterations in the design as he imagines necessary, repeating the calculations and making further alterations if requisite, until the result accords with his intention, and unless anything *very novel* or extraordinary is attempted this method will succeed pretty well, though of course, much will depend upon the proficiency of the Constructor, and the extent and quality of his precedents. Without these, the system is obviously insufficient.

CONSTRUCTION.

Wave system of J. Scott Russell. The system given in the greater part of this work, is familiarly known as the "wave" system of Mr. J. Scott Russell, and the shortest and best idea of it is as follows:

The genesis of the wave curves is as follows: The length of the fore-body as compared with the length of the after-body, is as 6 to 4, therefore the whole length is divided into ten equal parts, and six allotted to the fore-body. A circle whose diameter is equal to the half-breadth determined on, is described with its circumference touching the central line where the fore and after bodies join, its circumference is divided into sixteen equal parts, and the central lines of the fore and after bodies are each divided into eight equal parts; then for the curve of the fore-body from the foremost division on the central line, lay off the perpendicular distance of the central line from the first or lowest division on the circumference of the circle, and from the second division on the central line, the perpendicular distance of the second division on the circle, and so of each of the eight divisions; then through these points draw a line, and it will be the wave line curve forward. The curves of all the water-lines are similar.

Fore body.

After body. For the after-body lines, are drawn from the divisions on the circle parallel to the central line, on which the distances of the divisions on the central line from the fore-end of the after-body are respectively laid off from the divisions. A line drawn through these points will give the wave curve for the after-body.

"Chapman's" system, or parabolic system. But there is another system of Naval Construction, discovered by the celebrated Swedish Admiral, Chapman, and known as "the parabolic system;" and its applicability, completeness, and simplicity render it admirably adapted for every description of vessel.

Simplicity of Chapman's system. By it a Constructor can determine every particular of his vessel; he can be certain that she will have the required displacement, possess the proper amount of stability and trim as he intends, he has nothing to do, but after making a few preliminary calculations, to proceed to trace the vertical sections of the body plan. Nor do the advantages of the system stop here, as every variety of form, both of water-line and vertical section is equally applicable; in fact the Constructor has great latitude in the shape of his vessel, so long as he does not depart from the areas and centres of gravity settled upon at the outset. The outline of this method of construction is simply this:

Great latitude given to the constructor.

Origin of the "parabolic" system. Chapman endeavored to discover how far the areas of consecutive sections of the best known vessels followed any regular law. He, therefore, divided the area of each section by a constant quantity, *the breadth of the midship section*, and set off distances proportional to these quotients on perpendiculars to a base line, the perpendiculars being placed at intervals equal to the distance between the sections, and he found that the curve which passed through the ends of these perpendiculars, might be conveniently represented as a parabolical curve, both in the fore and after bodies, the vertex of the curve being at the middle of the base line, and the line representing the midship section forming the axis.

CONSTRUCTION. 129

Assuming then the fundamental equations *Chapman's fundamental equation.*

$$n = \frac{D}{l\,M - D} \quad (1)$$

$$k = (n + 2) \times a \quad (2)$$

$$y = p\,x^n \quad (3)$$

Where D is the load displacement in cubic feet,

<blockquote>
l is the length of the load water-line in feet.

a is the distance of the centre of gravity of displacement from the middle of the water-line in feet.

k is the distance of the midship section from the middle of the water-line in feet.

n being what is termed the exponent of the parabolic curve.

p the parameter.
</blockquote>

The two last quantities being used to assist the calculations; n is determined from equation (1), and p from the formula

$$p = \left\{ \frac{\frac{l}{2} \pm k}{M} \right\}^n \quad (4)$$

which is, in fact, a particular form of (3), and where, when the centre of gravity of displacement is abaft the middle of the water-line, the plus sign is used for the fore-body, the minus sign for the after-body, and *vice versa* when the centre of gravity of displacement is before the middle.

x and y being co-ordinates of the curves, or in other words y is the distance of any one section from the midship section, and x the difference between a line representing the area of the midship section, and a line representing the area of any one section.

From this last definition, it follows, that

<blockquote>
Area of section $= M - x$
</blockquote>

and from (3) $\quad x = \dfrac{y^n}{p} \quad (5)$

The Constructor has now the means of ascertaining the areas of the successive sections, by calculating n and p, and then substituting the successive values of y in the equation (5).

In practice the calculations are confined to one-half the vessel only, and therefore, instead of x and M, one-half of these values is taken; and with some degree of inaccuracy in the tables, as usually constructed, these halves are treated as whole areas.

For the sake of illustration, suppose the Constructor about to design a schooner yacht, *Example.*

Where *ft.*
Load displacement $= D = 5040$
Length of load water-line $= l = 80$
Area of midship section $= M = 110$

Distance of the centre of gravity of displacement from the middle $= a = 1\cdot 6$ ft. abaft.

17

130   CONSTRUCTION.

Example. Now to find n which $= \dfrac{D}{l\,M - D}$ (1)

or $= \dfrac{5040}{80 \times 110 - 5040} = 1.34$

Next, $k = (n + 2) \times a$ (2)
or $= (1.34 + 2)\,1.6 = 5.25$ abaft.

Then for the fore-body.

$\log. 2\,p = \log. 2 \left\{ \dfrac{\dfrac{l}{2} + k}{M} \right\}^{n} = \log. \left\{ \dfrac{\dfrac{80}{2} + 5.25}{55} \right\}^{1.34} = 478165 = \log. 3.$

For the after-body

$\log. 2\,p = \log. 2 \left\{ \dfrac{\dfrac{l}{2} - k}{M} \right\}^{n} = \log. \left\{ \dfrac{\dfrac{80}{2} - 5.25}{55} \right\}^{1.34} = 3245167 = \log 2.11$

As this quantity is used in multiplication only, the logarithm of it only need be found; then, by substituting the successive values of y in the equation

$$x = \dfrac{y^n}{p}$$

which must be calculated by logarithms, the following table is constructed:

For the fore-body $x = \dfrac{y^{\,1.34}}{3}$

y or distance from the midship section.	$\dfrac{x}{2}$ or abscissa.	Half area of M. S. $\dfrac{x}{2} - \dfrac{x}{2} =$ half area of section.
feet.	square feet.	square feet.
7.5	4.95	50.05
15.0	12.5	42.5
22.5	21.5	33.5
30.0	31.7	23.3
37.5	42.8	12.2

For the after-body $x = \dfrac{y^{\,1.34}}{2.11}$

y or distance from the midship section.	$\dfrac{x}{2}$ or abscissa.	Half area of M. S. $\dfrac{x}{2} - \dfrac{x}{2} =$ half area of section.
feet.	square feet.	square feet.
7.5	7.05	47.95
15.0	17.9	37.1
22.5	30.7	24.3
30.0	45.2	9.8

CONSTRUCTION.

A little careful study of the foregoing calculation will enable any one, with a slight knowledge of logarithms to comprehend the working of the system. No more calculations than the foregoing are required, as the position of the centre of gravity, the displacement and the area of the midship section, are all, by Chapman's method, pre-determined quantities, and form the basis of the design, therefore to calculate them after the design is completed, would be merely a useless repetition.

Before proceeding with the construction drawing of the schooner yacht taken as an example, it will be well to give the principal proportions which experience has shown to be useful and applicable to such vessels. *Data which experience has shown to be useful in the case here treated of.*

The breadth generally = the length $\times$ .26 for fast schooners.
The depth " = the breadth $\times$ .3952
The area of midship section = breadth $\times$ depth $\times$ .6.
" " " load water-line = breadth $\times$ length $\times$ .7021.

Load displacement in cubic feet = length $\times$ breadth $\times$ depth $\times$ .3623.

The midship section from the fore end of the water-line = length $\times$ .517.

The centre of gravity of displacement from the fore end of the water-line = length $\times$ .55 for schooners.

Or generally = length $\times$ .02 abaft the middle.

In making the construction drawing, the same order is to be observed as in copying, but the draughtsman is left to his own resources, as to the dimensions and forms of the different parts. *Caution to the Student.* His first care should be to understand distinctly and exactly the sort of vessel required. Let it be assumed, then, that this schooner yacht is to be 144 tons, or 5040 cubic feet the displacement required. The draft of water not to exceed eleven feet, and that she is to have as much speed as is consistent with a certain degree of accommodation. In Fincham or Marrett, the student will find tables giving the dimensions of some vessel of suitable character and *similar displacement*, and is thus enabled to fix the length of the water-line at 80 feet, then the breadth = .26 l = 20.8 feet.

But, in this case, as the breadth is rather limited, and accommodation is a desideratum, it may probably be better to give a small additional breadth to ensure sufficient stability, displacement and accommodation. Suppose, therefore, the breadth is determined at .27 l = 21.5 feet, then the depth = .3952 b = 8.5 feet.

If to this is added one foot for the depth of the keel below its rabbet, the mean draft of water is made 9.5 feet, and with the maximum draft of eleven feet aft, this gives eight feet for the draft of water forward. This, however, is rather more "drag" than may be advisable, and therefore, the draft forward may be increased to nine feet with advantage.

The load displacement = length × breadth × depth × .36
= 5,296 cubic feet = 151 tons, an excess of seven tons abo[ve]
that required, or

$$\frac{\text{displacement}}{\text{length} \times \text{breadth} \times \text{depth}} = \frac{5040}{14620} = .3447$$

or the proportion the displacement (5040 cubic feet,) bears [to]
the circumscribing parallelopipedon.

The Constructor will judge by comparison, whether this p[ro]portion is adapted to the required purpose, if not, some alte[ra]tion in the dimensions must be made.

The area of the midship section = breadth × depth × .6 [=]
110 square feet.

The exponent of the parabolic curve,

$$n = \frac{D}{l\,M - D} = \frac{5040}{80 \times 110 - 5040} = 1.34$$

Value of n in yacht 'America.' Which nearly corresponds with the value of n in the case [of] the yacht "America," and, therefore, it may be presumed t[hat] the proposed vessel will, by assigning that value to n, be of p[ro]portionate fullness in relation to the dimensions.

The distance of the centre of gravity of displacement ab[aft] the middle of the load water-line, will be length × .02 $l =$ [   ] feet $= a$, and the midship section will be from (2) 1.6 (1.34 + [   ]) $= 5.25$ feet *abaft* the middle.

The calculations for the areas of the sections of this ves[sel] have already been given, (page 130,) and therefore no repetit[ion] is here necessary.

Commencement of the drawing. Having arranged these preliminaries, the drawing may [be] commenced, the sheer, rake of the stern, and form of the count[er] can be taken from tables, or from other drawings, and altered [to] *Midship section.* suit the Constructor's judgment. The midship section in the sh[eer] plan, must be drawn at its proper distance from the fore end [of] the load water-line, and the other sections at the determi[ned] distances from the midship section; in the present case they [are] placed at intervals of seven feet six inches. In designing [the] midship section, which is done in the body plan, care must [be] taken that the half area = 55 square feet exactly; the sect[ion] being sketched in by the eye—its area may in this prelimin[ary] part of the work be found by the *"old rule"* thus:

Ordinates.

Load water-line,	10.75 (half.)
Second water-line,	10.1
Third water line,	7.0
Fourth water-line,	2.4
Keel,	. 3 (half.)

25.025
2.2 = distance between water-li[nes]

55.055

CONSTRUCTION.

It is hardly to be expected that, at the first trial, the midship section will be drawn of the correct area, but after one or two alterations it will generally be obtained. When this is done, the load water-line from the midship section to the fore end must be drawn in the half breadth plan. Having determined from Fincham or Marrett's tables of construction, or from other sources, the angle which it should make with the middle line forward, a line can be drawn for some distance from the ending at that angle; then, with a small penning batten, the breadth at the midship section is joined to this line; this will give a "straight" water-line. When any hollow is required in the water-line, the batten may be continued to the ending, describing the hollow required, or the load water-line may be formed according to Russell's wave system, if the Constructor prefers that system. Wave lines.

The half breadth lines of the deck and roughtree rail may next be drawn, of such shape as the Constructor thinks best. A section intermediate with the midship section and foremost extremity drawn in the body plan, (the half breadth of the load water-line being taken from the half breadth plan,) and altered, if necessary, until its half area corresponds with the area already determined for such section, will give a guide for drawing the other water-lines of the fore body in the half breadth plan. When the remaining sections are drawn, if their areas do not agree with the calculated areas, alterations must be made.

The after body is proceeded with in a similar manner, and when the whole of the sections are completed, the designer may, perhaps, require some alterations to be made.

When such alterations from the original plan, involve any considerable change of form or alteration of the several sectional areas, it may be advisable to calculate the displacement and the position of the centre of gravity of displacement, &c., in order to prevent too great a deviation from the original intention; but when no alteration, or at least only a slight one is made, this is not necessary. Plate II shows a design for a schooner yacht in conformity with the foregoing dimensions and calculations, and without any material alterations from the established areas, in order to show how well adapted for construction the parabolic system is.

To complete the vessel, the masts and sails have to be arranged. The area of the vertical longitudinal section $= l \times h = 80 \times 10 = 800$: area of load water-line $= l \times b \times .7 = 1204$, these sums multiplied together $= 963200$, which is the co-efficient for the dimensions of the spars. (See table III, pp. 36, 37, Marrett's yacht building.) Masts and sails.

The centre of effort should be placed at .006 of the length of load water-line $= .48$ feet abaft the centre of gravity of vertical longitudinal section, and at the height of the centre of effort, the main-mast will be one-tenth of the length of the water-line $= 8$ feet abaft the centre of the vertical longitudinal section. The fore mast will be .344 of the length of the water-line $= 27.5$ feet before the main-mast. Centre of effort of sail.

With these positions, a sail drawing as in Pl. III must be made to the dimensions given in Marrett's table of schooners' spars; and the area of sail and position of centre of effort, both as to height and length calculated. If the results do not agree with the established position for the centre of effort, some change must be made, either by re-adjusting the proportion of fore and after sail, or by moving the masts and preserving the same measurement of spars, as it is imperatively necessary that a proper and correct balance of sail should exist, otherwise, the care of the Constructor in designing the hull has been completely thrown away.

If the vessel is constructed on the wave theory, the mode of balancing the sail has already been shown.

CHAPTER XXVI.

MAKING THE CALCULATIONS.

The calculations for an ordinary vessel, are exceedingly simple —a little method and arrangement, however, shorten the process amazingly.

Two rules (generally known as Sterling's rules) are used by the majority of Constructors, they are briefly as follows:

Divide the base of the area, bounded by a curve, into any even number of equal parts, which of course, gives an uneven number of ordinates. The general expression of the rule is then

$$\text{Area} = [A + 4P + 2Q] \frac{r}{3}$$

Where A = sum of the first and last ordinates,

$4P$ = sum of the even ordinates multiplied by 4,

$2Q$ = sum of the remaining ordinates multiplied by 2,

and r = is equal to the linear measurement of the common interval between the ordinates.

The above formula may also be put in the following form:

$$\text{Area} = \left\{ \frac{A}{2} + 2P + Q \right\} \times \frac{2r}{3}$$

The second of Sterling's rules is as follows:

Divide the base of the area, bounded by a curve, into a number of equal intervals which shall be in number a multiple of three, which, of course, will give one additional to the number of ordinates, then the area of the figure may be determined by the following formula:

$$\text{Area} = \left\{ A + 2P + 3Q \right\} \times \frac{3r}{8}$$

where A = sum of the first and last ordinates,

$2P$ = sum of the 4th, 7th, 10th, 13th, &c., multiplied by 2,

$3Q$ = sum of the remaining ordinates, multiplied by 3 or 2nd, 3rd, 5th, 6th, 8th, 9th, &c. r is the common interval as before.

This formula may also be modified as follows:

$$\text{Area} = \left\{ \frac{A}{2} + P + 1.5\,Q \right\} \times \frac{3r}{4}$$

In the first rule, the curve bounding the area is supposed to be a portion of a common parabola, while, in the second rule, the curve is assumed to be a cubic parabola. The results from either rule, differ in so trifling a degree as to be practically insignificant.

The plan of calculation most applicable for a small vessel like the yacht in question, is as follows:

136   CALCULATIONS.

Plan of calculation, using a table to put the several quantities in.

On the drawing of the vessel, (Plate II,) fix No. 1 section, in the sheer plan at some determinate place, such as the fore end of the load water-line, then divide the length of that line into any uneven number of equal parts by lines perpendicular to it, which will represent the vertical sections. Draw a body plan of these sections to the outside of the plank, and having the load water-line as a base in the sheer plan. Divide the distance from the load water-line, to the line of the lower edge of the rabbet of the keel continued, at No. 1 section into an odd number of equi-distant parts, as five inclusive, also divide the corresponding distance at the aftermost section, into a like number of equi-distant parts, draw lines from No. 1 section to the aftermost section joining these divisions, and transfer the heights of the intersections of these lines with the sections, to the body plan. A table must now be ruled similar to the accompanying form, and the half breadth of each section at each water-line measured from the body plan, inserted in its proper place in the table. (*See Table, p.* 136½.)

A description of each column, is as follows:

a. The number of the vertical section.

Distance between the water-lines—how found.

b. The distance between the water-lines at each section respectively; it is thus found: the distance of the load water-line to the lower edge of the rabbet of the keel at No. 1 section, is eight feet, and at No. 11 section it is 9.875 feet, then as No. 1 and No. 11 sections are seventy-five feet apart, it follows that the difference of draft of water in seventy-five feet, is 1.875 feet, and the difference between each section when (as is in this case) they are 7.5 feet apart is

$$\frac{1.875 \times 7.5}{75} = .1875;$$

as there are four water-lines the distance between them, at each succeeding section will be increased by the fourth part of .1875 = .047 feet. At No. 1 section, the water-lines are two feet apart; at No. 2, 2 + .047 = 2.047 ft. apart, and so on.

c. The number by which the ordinates are to be multiplied in finding the displacement by horizontal sections.

d. The ordinates of the lowest longitudinal section, or half siding of the keel at each section.

e. The ordinate of the lowest water-line.

f. The ordinate of the third water-line.

g. The ordinate of the second water-line.

h. The ordinate of the load water-line.

i. The sums of the ordinates as multiplied.

k. The half areas of the sections $= \frac{i \times b}{3}$ thus, for No. 7 section, $\frac{72.2 \times 2.282}{3} = 54.9$ square feet.

l. The half areas of the sections multiplied by the numbers in column *c* respectively, the sum of the column *l* multiplied by one-third of the common interval between the sections, gives one-half the displacement in cubic feet, or $\frac{o\,l \times 7.5}{3} \times 2 =$ whole

CALCULATIONS FOR A SCHOONER YACHT OF EIGHTY FEET LENGTH.

a	b	c	d (keel)	e (wl.4)	f (wl.3)	g (wl.2)	h (lwl.1)	i	k	l	m	n	o	p	r	s	t	u	v	w
			1	4	2	4	1													
1	2·000	—	·3 / ·3	·3 / 1·2	·3 / ·6	·3 / 1·2	·3 / ·3	3·6	2·4	2·4			·3		·3	·3	·3	·3		·03
2	2·047	4	·3 / ·3	·6 / 2·4	1·4 / 2·8	2·1 / 8·4	2·8 / 2·8	16·7	11·4	45·6	1	45·6	11·2	11·2	2·8	2·5	2·6	10·4	10·4	21·95
3	2·094	2	·3 / ·3	1·3 / 5·2	2·7 / 5·4	4·1 / 16·4	5·5 / 5·5	32·8	22·9	45·8	2	91·6	11·0	22·0	5·9	4·8	5·4	10·8	21·6	166·37
4	2·141	4	·3 / ·3	1·8 / 7·2	4·1 / 8·2	6·1 / 24·4	7·8 / 7·8	47·9	34·2	136·8	3	410·4	31·2	93·6	8·4	7·0	7·7	30·8	92·4	474·55
5	2·188	2	·3 / ·3	2·0 / 8·0	5·1 / 10·2	7·8 / 31·2	9·4 / 9·4	59·1	43·1	86·2	4	344·8	18·8	75·2	9·8	8·7	9·2	18·4	73·6	830·50
6	2·235	4	·3 / ·3	2·2 / 8·8	6·1 / 12·2	9·4 / 37·6	10·3 / 10·3	69·2	51·5	206·0	5	1030·0	41·2	206·0	10·7	9·9	10·3	41·2	206·0	1092·73
7	2·282	2	·3 / ·3	2·3 / 9·2	6·4 / 12·8	9·8 / 39·2	10·7 / 10·7	72·2	54·9	109·8	6	658·8	21·4	128·4	11·0	10·4	10·7	21·4	128·4	1225·04
8	2·329	4	·3 / ·3	1·7 / 6·8	5·3 / 10·6	8·7 / 34·8	10·3 / 10·3	62·8	48·7	194·8	7	1383·6	41·2	288·4	10·8	9·7	10·3	41·2	288·4	1092·73
9	2·376	2	·3 / ·3	1·1 / 4·4	3·6 / 7·2	6·6 / 26·4	9·1 / 9·1	47·4	37·5	75·0	8	600·0	18·2	145·6	10·3	8·2	9·2	18·4	147·2	753·57
10	2·423	4	·3 / ·3	·8 / 3·2	1·9 / 3·8	4·0 / 16·0	7·0 / 7·0	30·3	24·5	98·0	9	882·0	28·0	252·0	8·7	5·8	7·2	28·8	259·2	343·00
11	2·470	—	·3 / ·3	·4 / 1·6	·7 / 1·4	1·2 / 4·8	3·2 / 3·2	11·3	9·3	9·3	10	93·0	3·2	32·0	4·4	2·5	3·5	3·5	35·0	32·77
		0	3·0	14·2	37·1	59·5	74·7		340·4	1009·7		5539·8	225·7	1254·4				225·2	1262·2	6033·24

Section No. I.—·25 feet abaft fore end of load-water line; distance between the vertical sections 7·5 feet.
Section No. II.—4·75 before after end of load-water line; keel below rabbit 1 foot.

CALCULATIONS.

displacement, which divided by 35, gives the displacement in tons of that part of the vessel included within the limits of the calculation, to this must be added the keel, rudder, &c., thus, *Distance between the water-lines—how found.*

$$1009.7 \times \frac{7.5}{3} \times 2 = 5048.5 \text{ cubic feet} = 144.2 \text{ tons.}$$

m. The multipliers for finding the distance of the centre of gravity of displacement from No. 1 section.

n. The products of $l \times m$, then $\frac{0\, n}{0\, l} \times$ common interval between the sections, = distance the centre of gravity of displacement is from No. 1 section or $\frac{5539.8}{1009.7} \times 7.5 = 41.2$ feet.

o. The ordinates of the load water-line multiplied by the numbers in column *c* respectively, the sum of these products multiplied by one-third of the common interval between the sections, will give half the area of the load water-line, thus:

$O\, o \times \frac{7.5}{3}$ —the half area of the load water-line, or $225.7 \times \frac{7.5}{3} \times 2 = 1128.5$ square feet = the whole area.

p. The products *m* and *o* for finding the distance of the centre of gravity of the load water-line from No 1 section. The sum of these products divided by the sum of column *o*, and the quotient multiplied by the common interval, gives the required distance, or $\frac{0\, p}{0\, o} \times 7.5$ = the distance of the centre of gravity of the load water-line from No. 1 section = $\frac{1254.4}{225.8} \times 7.5 = 41.68$ feet.

r. The ordinates of the *immersed* water-line when the vessel is inclined 10°.

s. The ordinates of the *emersed* water-line when the vessel is inclined 10°.

t. The sums of *r* and *s*.

u. The products of *t* and *c*. The sum of this column multiplied by one-third of the common interval, gives one-half of the area of the inclined water-line, thus

$O\, u \times \frac{7.5}{3} \times 2$ = the whole area, or $225.2 \times \frac{7.5}{3} \times 2 = 1126$ square feet.

v. The products of *u* and *m* for finding the distance of the centre of gravity of the inclined water-line from No. 1 section, then $\frac{0\, v}{0\, u} \times 7.5$ = distance required, or $\frac{1262.2}{225.2} \times 7.5 = 42.03$ feet.

w. The cubes of the ordinates of the load water-line required for finding the height of the meta-centre above the centre of gravity of displacement. The sum of this column multiplied by two-thirds of the common interval, and the product divided by the number of cubic feet in the displacement, gives the height of the meta-centre, thus $\dfrac{6033\ 24}{5048.5} \times \dfrac{7.5}{3} \times 2 = 5.975 =$ height of the meta-centre above the centre of gravity of the displacement.

Volume of displacement of fore-body. — To find that part of the displacement which is before the centre of gravity, this latter point being 41.2 feet abaft No. 1 section, it will be 3.8 feet before No. 7. The displacement between Nos. 1 and 7 taken from column l, is 2884.5 cubic feet, from this must be subtracted the cubical content of that part between No. 7 and the centre of gravity = area of No. 7 $\times$ 3.8 = 109.8 $\times$ 3.8 = 417 cubic feet; thus making the displacement before the centre of gravity 2467.5 cubic feet; and the displacement abaft that centre will be the whole displacement *minus* 2467.5 = 2581 cubic feet.

Volume of displacement of after-body.

Centre of gravity of fore body from centre of buoyancy. — The distance of the centre of gravity of the fore-body from the centre of gravity of displacement, will be for the part between Nos. 1 and 7, from column n

$$
\begin{array}{rl}
2.4 \times 0 = & \\
45.6 \times 1 = & 45.6 \\
45.8 \times 2 = & 91.6 \\
136.8 \times 3 = & 410.4 \\
86.2 \times 4 = & 344.8 \\
206.0 \times 5 = & 1030.0 \\
54.9 \times 6 = & 329.4 \\
\hline
576.9 & 2251.8
\end{array}
$$

$\dfrac{2251.8}{576.9} = 3.9$ which multiplied by the common interval 7.5 = 29.25 feet = distance the centre of gravity of that part between Nos. 1 and 7 is from No. 1 section. The centre of gravity of the part between No. 7 and the centre of gravity of displacement will be $45 - \dfrac{3.8}{2} = 43.1$ feet from No. 1, and combining these distances we have

Cubic contents.	Movements.
2884.5 × 29.25 =	84371
417.0 × 43.1 =	17972
2467.5	66399

$\dfrac{63399}{2467.5} = 26.9$ feet the distance of the centre of gravity of the fore-body from No. 1; then 41.2 — 26.9 = 13.9 feet, which will be the distance of the centre of gravity of the fore-body from the centre of gravity of displacement.

Centre of gravity of after-body from centre of buoyancy. — For the after-body, a similar method is pursued, except that in this case the smaller moment is positive. The result gives the distance as 13.5 feet.

CALCULATIONS.

To find the distance the centre of gravity of displacement is below the load water-line, multiply the sums of the products of the ordinates of the water-lines multiplied by the numbers 1, 4, 2, 4, 1 respectively by 0, 1, 2, 3, 4 respectively, commencing from the load water-line downwards; the sum of these products divided by the sum of the first products, and the quotient multiplied by the distance between the water-lines at the centre of gravity of displacement will give the required distance, thus:

To find the position of centre of buoyancy.

Load water-line	$74.7 \times 1 =$	74.7	$\times 0$	
Water-line (2)	$59.5 \times 4 =$	238.0	$\times 1 =$	238.0
" (3)	$37.1 \times 2 =$	74.2	$\times 2 =$	148.4
" (4)	$14.2 \times 4 =$	56.8	$\times 3 =$	170.4
Keel	$3.0 \times 1 =$	3.0	$\times 4 =$	12.0
		446.7		568.8

The distance between the water-lines at the centre of gravity is thus found; *as* the length between the extreme sections *is to* the difference of depth in that length, *so is* the length from No. 1 section, to the centre of gravity *to the* difference in depth due to that part of the length; or,

feet. feet. feet. feet.
80 : 2 : : 41.2 : 1.03

this added to the depth at No. 1 section, gives the whole depth at the centre of gravity $= 8 + 1.03 = 9.03$ feet, which divided by 4, (the number of water-lines,) $=$ the distance the water-lines are apart at the centre of gravity $= 2.26$ feet.

Then $\dfrac{568.8}{446.7} \times 2.26 = 2.877$ feet the required distance.

This method of finding the height of the centre of gravity of each water-line, is not always strictly correct, because the centre of gravity of displacement longitudinally, may not be at the centre of gravity of displacement, but the error is so trifling as to be practically insignificant. In the foregoing calculations the keel is omitted, its dimensions are, length 80 feet, depth 1 foot, and breadth 7 inches, the cubic content is therefore 48 feet.

There is also a trapezium between No. 2 section and the stern-post, the depth 10 feet, length 5 feet, and mean breadth 1 foot $=$ 50 cubic feet; the total displacement is therefore

$5048.5 + 48 + 50 = 5146.5$ cubic feet $= 147$ tons.

A full understanding of the reason for each step in the process of the calculations, greatly facilitates the proceeding.

Reasons for the foregoing system.

The first and principal object is to find the area of each vertical section. If the section were a rectilinear triangle, nothing more would be required than to take the sum of half of the two extreme ordinates and multiply it by its depth, but the sections of a ship have one side of the triangle curvilinear, and therefore this easy method of finding the area, would give an erroneous result, as it does not include the space contained between the right line and the curve.

There are three methods of obtaining a more correct area, the rule for the first method has not been given, as although it is the simplest known, it is too inaccurate for most cases.

140 CALCULATIONS.

"Short" rule. This rule is:—Divide the section into a number of equi-distant ordinates, add half the sum of the first and last ordinates to the sum of the remaining ordinates, and multiply their sum by the distance between them, applying this rule to section 7 of the table, we have:

Ordinate 1 — .3	Ordinate 2 — 2.3
" 5 — 10.7	" 3 — 6.4
2] 11.0	" 4 — 9.8
5.5	18.5
18.5	
24.0	

2.282 distance between the ordinates.

54.768 required half area.

Which is slightly different from the tabulated area.

1st of Sterling's rules generally known as 'Chapman's' rule. But it is better to use Sterling's rules given at the commencement of this chapter. The first of these is sometimes called "Chapman's" rule—the second, the "three-eights" rule.

By the first we would have

Ordinate 1 —	0.3	Ordinate 2 —	2.3	Ordinate 3 —	6.4
" 5 —	10.7	" 4 —	9.8		2
A.	11.0	P.	12.1	2 Q.	12.8
	48.4		4		
	12.8	4 P.	48.4		
	72.2				

r 2.282 the distance between the ordinates.

3) 164.76

54.92 the required half area.

as determined from rule area $= [A + 4P + 2Q] \dfrac{r}{3}$

the curve bounding the area being supposed to be a portion of a common parabola.

2nd as the "three eights rule," or "three plus one rule." The second rule may be applied in a similar manner, but the area must be divided into that number of equal divisions, which will be a multiple of 3. So as to make the number of ordinates a multiple of 3 with one added.

Which rule may be used. As a general rule in calculating areas or displacement, when the water-lines or sections are very round, or when the vertical sections are more than five feet apart, either Chapman's rule or the "three-eights" rule should be employed, but when the lines are tolerably straight, and the ordinates are less than five feet apart, the *"short rule"* will give a result sufficiently accurate, and from its greater simplicity is to be preferred, or the two rules may be combined with advantage, as when the vertical sections are

CALCULATIONS. 141

much curved, but the water-lines rather straight, thus, to calculate the displacement, taking the half areas found by Chapman's rule from column k, page 136½. *Calculating Displacement.*

Section 1. 2.4 (half.) *By vertical sections.*
" 2. 11.4
" 3. 22.9
" 4. 34.2
" 5. 43.1
" 6. 51.5
" 7. 54.9
" 8. 48.7
" 9. 37.5
" 10. 24.5
" 11. 9.3 (half.)
———
334.55
7.5 distance between the sections.
———
2509.125 half displacement.
2
———
5018.25 whole displacement.

When the water-lines are parallel and equi-distant, the displacement may be found by taking the areas of the water-lines and applying either of the rules to them; in the case of the yacht whose displacement we are calculating, the water-lines not being equi-distant and parallel, it is almost impracticable to obtain an exact result in this way; though a tolerably good approximation may be obtained as in the following example, the distance between the water-lines being supposed to be that at the middle of the length, or at No. 6 section, then taking the areas of the water-lines by the "short" rule, and completing the calculation by Chapman's rule we have *By horizontal sections.*

Keel   3.0 × 1 =   3.0
Water line (4)   14.2 × 4 = 56.8
"   " (3)   37.1 × 2 = 74.2
"   " (2)   59.5 × 4 = 238.0
Load water-line   74.7 × 1 = 74.7
                              ———
                              446.7
                              7.5 common interval.
                              ———
                         3) 3350.25
                              ———
                              1116.75
                              2.235 { distance between the water-lines.
                              ———
                              2495.9 half displacement.
                              2
                              ———
                              4991.8 whole displacement.

which differs very little from the more laborious process, and if not quite so correct, is useful as a *check* on the *accuracy* of the work.

142   CALCULATIONS.

Finding the position of centre of buoyancy.   In calculating the distance of the centre of gravity of displacement from No. 1 section, it is required to find the moment of each section, which is its distance multiplied by its contents, the sum of these moments divided by the sum of the contents will give the distance required, but as this process involves large sums of figures, it is curtailed by multiplying the area of each section by the number of its place, from No. 1, if then the sum of the areas so multiplied is divided by the sum of all the areas, and the quotient is multiplied by the distance between the sections, the same end is gained with a less amount of figuring.

Centres of gravity.   The position of the other centres of gravity are calculated on the same principle.

Height of meta-centre above centre of buoyancy.   The height of the meta-centre above the centre of gravity of displacement, being a measure of the comparative stability of the ship, is estimated from the expression $\dfrac{2}{3} \S \dfrac{y^3 \, dx}{D}$, in which

   y   = the ordinates of the half breadth load water section.
   d x = the increment of the length of load water section.
   D   = Displacement in cubic feet.

The ordinates are taken from the table, column *w*, and cubed, and the calculation made in accordance with the rules previously given, *pages* 137, 138.

Calculations simple and easy of performance.   It will thus be seen that the calculations for a small vessel are extremely simple and easy of performance, indeed any one tolerably well versed in simple arithmetic may complete the whole of the calculations for a vessel of, say 200 tons *in less than an hour*, when equi-distant vertical sections and the water-lines at their proper height are drawn in the half breadth and body plans.

To find area of sail and centre of effort.   It now only remains to find the area and position of the centre of effort of the sails.

The area of triangular sails like the jib, is found as follows:

Area of triangular sails.   Make a sketch (by scale) of the sail on paper, the *luff* being the hypothenuse, the *foot* the base, the *after leech* the perpendicular, then from the clew let fall a perpendicular to the luff, multiply the length of luff by length of this perpendicular and divide by 2, the result is the required area.

Centre of gravity of a triangular sail.   The centre of gravity (or of effort) of any triangular sail like the jib, is found by bi-secting the length on the luff, then two-thirds the distance from this point to the clew, set off on a straight line, is the position of the centre of gravity required.

Area of a trapezium.   When the shape of the sail is a trapezium, like the mainsail of a schooner, the area is ascertained by dividing the whole plane into two triangles by a diagonal, finding the area of each triangle, and adding them together.

To find the centre of gravity of same.   To find the centre of gravity of the same sail, divide it by lines into four triangles, find the centre of gravity of each, then draw lines from these centres—where they intersect, is the centre of gravity of the whole sail.

CALCULATIONS.

The area of the sails of the schooner (Plate III,) calculated in the foregoing manner will be as follows: *Area of sails of schooner taken as an example.*

	Areas in sq. feet.
Jib,	992
Foresail,	1290
Mainsail,	2028
	4310

The position of the centre of effort as to height above the water-line, is found by multiplying the area of each sail, by the perpendicular distance of its centre of gravity from the waterline, then dividing the sum of these products or moments by the sum of the areas, the quotient is the required distance, thus: *Height of centre of effort above load water line.*

Areas in sq. feet.	Height of C. of G. above L. W. line in feet.	Momenta.
Jib, 992	× 25.6	= 25395·
Fore sail, 1290	× 32.0	= 41280·
Main sail, 2028	× 31.6	= 64084·
4310		) 130719
		30.33

or height of the centre of effort above the load water-line.

The distance of the centre of effort from any perpendicular to the water-line, considered as an initial point, is found by dividing the difference of the moment of sail before and abaft that perpendicular by the area of sail; according as the excess of the momenta is before or abaft the perpendicular, so will the position of the centre of effort be; thus: *Position of centre of effort finally determined.*

	Areas, square feet.	Distance of the centre of gravity from a perpendicular at the centre of the longitudinal section in feet.	Momenta.
Jib,	992 ×	35.1 before	= — 34819
Fore sail,	1290 ×	6.0 before	= — 7740
Main sail,	2028 ×	27.5 abaft	= + 55770
	4310		

or [34819 + 7740 — 55770] ÷ 4310 = 3.06, which is the

distance the centre of effort of all the sails is *abaft* the centre of gravity of the immersed longitudinal section.

CHAPTER XXVII.

HOW TO SET ABOUT THE DESIGN OF A MAN-OF-WAR.

Example of design of a man-of-war.

When a merchant ship is spoken of, it is usual to say a vessel of so many tons, indicating the vessel's capacity and the weight she can carry; but when a man-of-war is spoken of, it is usual to designate her by the number of guns she carries, meaning, of course, her weight of metal.

A Naval Architect may be able to design a merchant vessel to perfection, and yet be totally unable to design a man-of-war; especially with the modern ships is this a task most difficult, they carry a quantity of heavy iron plates in such a position that, when no special care is bestowed upon the design, the vessel must prove a perfect failure. In a merchant vessel the weights are carried below, while in a man-of-war they are nearly all carried above the water-line.

Suppose a man-of-war is to be designed to carry 50 guns, and of these 36 upon the main deck and 14 on the spar deck, the latter being distributed on the forecastle and quarter deck, while the main deck guns are distributed 18 in each broadside. It is the length of the ship that is mainly affected by the 36 broadside guns. Experience has proved that the lower port sill must be at least 9 feet out of the water, but in some vessels, that height has been increased to 10 and even to 11 feet. Suppose the height determined to be 10 ft. 6 inches, the ports to be 3 feet high and a space of 2 feet kept above the upper port sill, then the Constructor has already a height out of water of 10.5 ft. $+ 3 + 2 = 15$ ft. 6 inches.

This will be the height out of water of the top of the beam of the upper deck at the side of the ship. He has, therefore, already made a side wall for the vessel's battery of, say, 16 feet out of water. Now what should the *length* of this side wall be? In former days the space between the guns was seldom more than 8 feet. The distance now-a-days, on account of lateral train, is extended to 15 and even 18 feet,* but assume 15 feet as a distance from centre to centre. The Constructor has therefore, for the length to be occupied by his battery, $18 \times 15 = 270$ feet.

Suppose further, that the speed of the vessel is to be 13 *knots*; to this speed belongs a length of ship of 160 feet, or a length of *entrance* of 94 feet, and a length of *run* of 67 feet.† Make the length of the ship then, say 300 feet, and assume the beam to be 52 feet, or a proportion of 6 to 1. The Constructor has then the following preliminary dimensions:

Dimensions.

Length 300 feet.
Breadth 52 "
Height of lower port sill above L. W. line, 10 ft. six inches.

Draft of water.

Next comes the element of draft. A man-of-war must always be supposed in a loaded condition, or down to her load water-

---

* For XI inch gun on iron carriage, 18 feet.
† Supposing that she is designed according to the "Wave" system. See Table, p. 99.

DESIGN OF A MAN-OF-WAR.   145

line, that water-line being the line on which she has got to do the work. The light water-line may, therefore, be left out of the question. Now if this man-of-war were to be of box form or of rectangular shape with square bilges, the table on page 32 would give the critical proportions of draft to breadth, for by that table it is seen that to a beam of 54 feet belongs a critical draft of 22 feet, and that such a form would not be able to carry top-weight. The adopted beam is, however, 52 feet, and therefore the Constructor may safely assume a draft of 22 feet, especially as he intends probably to give *rise of floor* to the midship section. Suppose, further, that the co-efficient of fineness of the midship section is 0.8: the area of the midship section will then be $22 \times 52 \times .8 = 915.2$ square feet. Now is the time for the designer to see in how far he can adopt fine lines and still have displacement enough to carry the weight required.

For a speed of 13 knots, it was seen that a length of entrance and run was needed of 161 feet, which, subtracted from 300 ft., leaves a length of middle body of 139 feet. Hence the Constructor has for the displacement,

Middle body, 127,212.8 cubic feet.   *Displacement*
The ends,   87,840.0   "
_____
Total,   215,052.8   "

But the Constructor needs $6,300 \times 35 = 220,500$ cubic feet. He has, therefore, $220,500 - 215,052.8 = 5447.2$ cubic feet, or 155.6 tons too little.

To make this up, the midship section may be made fuller without damage; say 950 square feet nearly. The displacement is thus increased to 224,000 cubic feet; more than enough; and then, taking the fraction of 0.6 for the fineness of the ends, the Constructor has the following:

Length on load water-line,	300 feet.	*Quantities.*
" of entrance,	94 "	
" " run,	67 "	
" " straight middle body,	139 "	
Beam,	52 "	
Draft of water,	22 "	
Depth at the side,	42 "	

Area of immersed midship section, $= 22 \times 52 \times .83 = 949.52$ square feet.

Displacement of middle body 131,983 cubic feet, or 3771 tons.

Displacement of fore and after-bodies together 91,770 cubic feet, or 2622 tons, or

	131,983 cubic feet		$=$	3771 tons.
	91,770	"	" $=$	2622 "
Total,	223,753	"	"	6393 "
Required,	220,500	"	"	6300 "
Surplus,	3,253	"	"	93 "

19

DESIGN OF A MAN-OF-WAR.

Stability.

This surplus might easily be rectified by making the lines finer, or, in other words, by taking a little away from the middle body and making longer ends. But the Constructor may be satisfied that with the given dimensions and proportions, the ship will carry her weights.

Now comes the question of stability, for which (following Russell's mode of treating the subject) he obtains the following:

Volume of shoulder of the ends $= 0.1875 \times b \times L \times c + 0.3927 \times b^2 \times c$, wherein $b = 26$ $L = 161$, and $c = 13$; hence height of meta-centre above L. W. line $= 1.992$ feet.

Taking approximately the centre of displacement as 10 feet below the load water-line, the interval of 11.992 or nearly 12 feet is obtained. This may seem small, but it must be considered that the power of the shoulder is taken at its minimum, and, in reality, may be increased from 25 to 50 per cent., while the bottom buoyancy is maintained.

Weights.

The next point is to see the effect the weights will have.

The guns are to weigh say 500 tons, and the centre of gravity of this weight is to be, say 13.5 feet above the L. W. line. The centre of gravity of the hull alone may be supposed to be in the load water-line, which should be the case in men-of-war, their hulls generally being as much out of water as in it. Further, suppose the centre of weight of the engines, boilers, coals, &c., to be situated at half the draft, or 11 feet below the water-line. To find, therefore, the common centre of gravity, or the centre of all the weights, the

Moment of guns $= 500 \times 13.5 = 6750$ foot tons,
" of masts, spars, &c., $= 155 \times 80 = 12,400$ foot tons.

Considering these moments as negative, their sum is 19,150 tons; then,

Moment of hull, $= 3000 \times 0$
" of engines, boilers, &c., $= 1000 \times 11 = 11,000$ foot tons.
" of coals, $= 1000 \times 11 = 11,000$ foot tons.
" of sundries, $= 738 \times 10 = 7380$ foot tons.

Considering these last as positive, their sum is 29,380; deducting from this the negative moments, there remains $29,380 - 19,150 = 10,230$ foot tons, and this divided by the sum of all the weights, gives the place of the common centre of gravity as
$$\frac{10,230}{6393} = 1.6 \text{ feet below the load water line.}$$

The distance of the centre of all the weights from the meta-centre, will, therefore, be equal to $1.992 + 1.6 = 3.592$ feet. The moment of stability with weight, at an angle of 14° 2', is therefore $3.592 \times 6393 \times 0.2425$, or 5568.4 foot tons.

Sail area.

The next point is to determine the sail area. In merchant ships six times the longitudinal area is taken; in sailing yachts 6 to 12 times that area, but in men-of-war, from 6 to 4 times the

DESIGN OF A MAN-OF-WAR. 147

area is usual, the old frigates having, generally, but 4 times the area of longitudinal immersed section.

Assume the sail area to be five times the longitudinal area, or $5 \times 300 \times 22 = 33,000$ square feet of canvas, or taking the proportion of 36 square feet of canvas to every square foot of midship section, the sail area would be 34,182 square feet. The smaller quantity, however, is better, and this will give a length of main-yard of 90 feet—hence 88 feet is the height of the centre of effort above the water-line.

Supposing the pressure on the sails to be equal to 1 lb. to the square foot, it results that there is 34,182 lbs. acting with a leverage of 88 square feet; but since the vessel has careened 14° 2′, the cosine of this angle must be introduced as a factor. This gives 1291.26 foot tons as the upsetting moment due to the sails. Now the moment of stability with weight being 5568.4 foot tons, it is seen that there is surplus stability sufficient to bear a pressure of $4\frac{1}{2}$ lbs. instead of 1 lb. on each foot of standard sail area, which is a margin more than sufficient. *Stability under sail.*

The sail area, therefore, might be, if necessary, much increased, the ship having ample stability.

The next question is, whether the engines will drive the ship 13 knots per hour? For this purpose, the Constructor must see what head resistance is to be overcome. *Engine power necessary.*

By table on page 100, it will be seen that to a speed of 13 knots, belongs a head resistance of 482 lbs. to every square foot of midship section. The length of bow gives for a diminished fraction $(52-94) = .306$. The Constructor, therefore, has for head resistance $949.52 \times 3.06 \times 482 = 140,046$ lbs.

Horse power necessary to overcome the above equal to $19.2 \times .306 \times 949.52 = 5578.62$ indicated horse power.

For the wet surface, the following is a near approximation:

Periphery of midship section,	80 feet.
Skin of middle body,	$80 \times 139.0 = 11,120$ square feet.
Skin of fore and after bodies,	$80 \times 161 \times .5 = 6,440$ " "
Total =	17,560 " "

or, at 1 lb. equal to 17,560 lbs.

Horse power necessary to overcome the last named element of resistance, is

$$\frac{17,560}{482} \times 19.25 = 701.28$$ indicated horse power. The total power required is, therefore, $5578.62 + 701.28 = 6279.90$.

Suppose the engines to have worked up to five times their nominal power, this nominal power would then be equal to 1255.98 horse power. To this must be added one-fifth for *slip, and the power consumed by the engines themselves,* and the Constructor finally gets for the power required to do the work 1507.17 nominal, or 7535.85 indicated horse power.

DESIGN OF A MAN-OF-WAR.

Hence there is for the man-of-war in question, the following principal dimensions:

Dimensions.

Length on L. W. line,	300 feet.
" of entrance,	94 "
" " run,	67 "
" " middle body,	139 "
Breadth, extreme,	52 "
Depth at side,	42 "
Draft of water,	22 "
Tonnage, builder's,	3866 tons.
Displacement,	6393 "
Speed, in knots,	13
Area of immersed midship section,	949 sq. ft.
Distance between the ports,	15 feet.
Height of lower port-sill above L. W. line,	10 ft. 6 in.
Number of guns,	50
Indicated horse-power required,	7535·85

From which the drawings are made, and the ship constructed.

Nav. Arch.

APPENDIX.

An Explanation of the Terms, and of some Elementary Principles, Requisite to be Understood in the Theory and Practice of Ship-building.

AFLOAT.—Borne up by, or floating in, the water.

AFTER-BODY.—That part of a ship's body abaft the midships or dead-flat. This term is more particularly used in expressing the *figure* or *shape* of that part of the ship.

AFTER TIMBERS.—All those timbers abaft the MIDSHIPS or DEAD FLAT.

AIR FUNNEL—A cavity framed between the sides of some timbers, to admit fresh air into the ship, and convey the foul air out of it.

AMIDSHIPS.—In midships, or in the middle of the ship, either with regard to her length or breadth. Hence that timber or frame which has the greatest breadth and capacity in the ship is denominated the *midship bend*.

ANCHOR-LINING.—The short pieces of plank, or of board, fastened to the sides of the ship, or to stantions under the fore channel, to prevent the bill of the anchor from wounding the ship's side, when fishing the anchor.

TO ANCHOR STOCK.—To work planks in a manner resembling the stocks of anchors, by fashioning them in a tapering form from the middle, and working or fixing them over each other, so that the broad or middle part of one plank shall be immediately above or below the butts or ends of two others. This method, as it occasions a great consumption of wood, is only used where particular strength is required, as in the SPIRKETTINGS under ports, &c.

AN-END.—The position of any mast, &c., when erected perpendicularly on the deck. The top-masts are said to be AN-END when they are hoisted up to their usual stations.—This is also a common phrase for expressing the driving of any thing in the direction of its length, as to force one plank, &c., to meet the butt of another.

ANGLE OF INCIDENCE.—The angle made with the line of direction, by an impinging body, at the point of impact; as that formed by the direction of the wind upon the sails, or of the water upon the rudder, of a ship.

APRON.—A kind of false or inner stem, fayed on the aftside of the stem, from the head down to the dead-wood, in order to strengthen it. It is immediately above the foremost end of the keel, and conforms exactly to the shape of the stem, so that the convexity of one applied to the concavity of the other, forms one solid piece, which adds strength to the stem, and more firmly connects it with the keel.

ARCH OF THE COVE.—An elliptical moulding sprung over the cove at the lower part of the taffarel.

BACK OF THE POST.—The after-face of the STERN POST.

BACKSTAY STOOL.—A short piece of broad plank, bolted edge-ways to the ship's side, in the range of the channels, to project, and for the security of, the dead-eyes and chains for the Backstays. Sometimes the channels are left long enough to answer the purpose.

BACK-SWEEP.—*See* FRAMES.

BALANCE FRAMES.—Those *frames*, or bends of timber, of the same capacity or area, which are equally distant from the centre of gravity. *See* FRAMES.

BALLAST.—A quantity of iron, stone, or gravel, or such other like materials, deposited in a ship's hold, when she has no cargo, or too little to bring her sufficiently low in the water. It is used to counterbalance the effort of the wind upon the sails, and give the ship a proper stability, that she may be enabled to carry sail, without danger of over-setting.

BARK.—A name given to ships having three masts without a mizen top sail.

BARREL.—The main piece of a capstan or steering wheel. *See* CAPSTAN.

BATTENS.—In general, light scantlings of wood. In ship-building, long narrow laths of fir, their ends corresponding and fitted into each other with mortice and tenon, used in setting fair the sheer-lines on a ship. They are painted black in order to be the more conspicuous. Battens used on the mold-loft floor, are narrow laths, of which some are accurately graduated and marked with feet, inches, and quarters, for setting off distances. Battens for gratings are narrow thin laths of Oak.

BEAMS.—The substantial pieces of timber which stretch across the ship, from side to side, to support the decks and keep the ship together by means of the *Knees*, &c.; their ends being lodged on the clamps, keeping the ship to her breadth.

BEAM ARM, OR FORK BEAM, is a curved piece of timber, nearly of the depth of the beam, scarphed, tabled, and bolted, for additional security to the sides of beams athwart large openings in the decks, as the main hatchway and the mast-rooms.

BREAST BEAMS are those beams at the fore-part of the quarter deck and poop, and after part of the forecastle. They are sided larger than the rest; as they have an ornamental rail in the front, formed from the solid, and a rabbet one inch broader than its depth, which must be sufficient to bury the deals of the deck, and one inch above for a spurn-water. To prevent splitting the beam in the rabbet, the nails of the deck should be crossed, or so placed, alternately, as to form a sort of zigzag line.

HALF-BEAMS are short beams introduced to support the deck where there is no framing, as in those places where the beams are kept asunder by hatchways, ladder ways, &c. They are let down on the clamp at the side, and near midships into fore and aft carlings. On some decks they are, abaft the mizen-mast, generally of fir, let into the side tier of carlings.

THE MIDSHIP BEAM is the longest beam of the ship, lodged in the midship-frame or between the widest frame of timbers.

BEARDING.—The diminishing of the edge or surface of a piece of timber, &c., from a given line, as on the dead-wood, clamps, plank-sheers, fife-rails, &c.

BEARDING-LINE.—A curved line occasioned by bearding the dead-wood to the form of the body; the former being sided sufficiently, this line is carried high enough to prevent the heels of timbers from running to a sharp edge, and forms a rabbet for the timbers to step on; hence it is often called the STEPPING LINE.

BED.—A solid framing of timber to receive and to support the mortar in a Bomb Vessel.

BEETLE.—A large mallet used by Caulkers for driving in their reeming irons to open the seams, in order for caulking.

BELLY.—The inside or hollow part of compass or curved timber, the outside of which is called the BACK.

BELL-TOP.—A term applied to the top of a quarter gallery when the upper stool is hollowed away, or made like a rim, to give more height, as in the quarter galleries of small vessels, and the stool of the upper finishing comes home to the side, to complete overhead.

BEND-MOULD, in whole moulding. A mould made to form the futtocks in the square body, assisted by the rising-square, and floor-hollow.

BENDS.—The frames or ribs that form the ship's body from the keel to the top of the side at any particular station. They are first put together on the ground. That at the broadest part of the ship is denominated the MIDSHIP-BEND or DEAD-FLAT. The fore part of the Wales are commonly called *Bends*.

BETWEEN-DECKS.—The space contained between any two decks of a ship.

BEVEL.—A well known instrument, composed of a stock and a moveable tongue, for taking of angles on wood, &c., by shipwrights called BEVELLINGS.

BEVELLING BOARD.—A piece of deal on which the bevellings or angles of the timbers, &c., are described.

BEVELLINGS.—The windings or angles of the timbers, &c., a term applied to any deviation from a square or right-angle. Of Bevellings there are two sorts, denominated

Standing Bevellings and *Under Bevellings*. By the former is meant an obtuse angle or that which is *without a square;* and, by the latter, is understood an acute angle or that which is *within a square.*

BILGE.—That part of a ship's floor, on either side of the keel, which has more of an horizontal than of a perpendicular direction, and on which the ship would rest if laid on the ground: or, more particularly, those projecting parts of the bottom which are opposite to the heads of the floor-timbers amidships, on each side of the keel.

BILGE TREES, OR BILGE PIECES, OR BILGE KEELS.—The pieces of timber, fastened under the bilge of boats or other vessels, to keep them upright when on shore, or to prevent their falling to leeward when sailing.

BILGEWAYS.—A square bed of timber, placed under the bilge of the ship, to support her while launching.

BINDINGS.—The iron links which surround the *Dead Eyes.*

BINDING STRAKES.—Two strakes of oak plank, worked all the way fore and aft upon the beams of each deck, within one strake of the coamings of the main hatchway, in order to strengthen the deck, as that strake and the midship strakes are cut off by the pumps, &c.

BINS.—A sort of large chests, or erections in store-rooms, in which the stores are deposited. They are generally 3 or 4 feet deep, and nearly of the same breadth.

TO BIRTH-UP.—A term generally used for working up a topside or bulk-head with board or thin plank.

BLACK-STRAKE.—A broad strake, which is parallel to, and worked upon the upper edge of, the Wales, in order to strengthen the ship. It derives its name from being paid with pitch, and is the boundary for the painting of the topsides. Ships having no ports near the Wales, have generally two black-strakes.

BLOCKS FOR BUILDING THE SHIP UPON, are those solid pieces of oak timber fixed under the ship's keel, upon the groundways.

BOARD.—Timber sawed to a less thickness than plank; all broad stuff of or under one inch and a half in thickness.

BODIES.—The figure of a ship, abstractedly considered, is supposed to be divided into different parts, or figures, to each of which is given the appellation of *Body.* Hence we have the terms FORE BODY, AFTER BODY, CANT BODIES, and SQUARE BODY. Thus the *Fore Body* is the figure, or imaginary figure, of that part of the ship afore the midships or dead-flat, as seen from ahead. The *After Body*, in like manner, is the figure of that part of the ship abaft the midships, or dead-flat, as seen from astern. The *Cant Bodies* are distinguished into *Fore* and *After*, and signify the figure of that part of a ship's body or timbers, as seen from either side, which form the shape forward and aft, and whose planes make obtuse angles with the midship line of the ship; those in the Fore Cant Body being inclined to the stem, as those in the After one are to the stern post. The *Square Body* comprehends all the timbers whose areas or planes are perpendicular to the keel and square with the middle line of the ship; which is all that portion of a ship between the Cant Bodies.

BOLSTERS,—Pieces of oak timber fayed to the curvature of the bow, under the Hawse-Holes and down upon the upper or lower cheek, to prevent the cable from rubbing against the cheek.

BOLSTERS FOR THE ANCHOR LINING, are solid pieces of oak, bolted to the ship's side, at the fore part of the fore chains, on which the stantions are fixed that receive the anchor lining. The fore end of the bolster should extend two feet or more before the lining, for the convenience of a man's standing to assist in fishing the anchor.

BOLSTERS FOR SHEETS, TACKS, &c., are small pieces of fir or oak fayed under the Gunwale, &c., with the outer surface rounded to prevent the sheets and other rigging from chafing.

BOLTS.—Cylindrical or square pins of iron or copper, of various forms, for fastening and securing the different parts of the ship, the guns, &c. The figure of those for fastening the timbers, planks, hooks, knees, crutches, and other articles of a similar nature, is cylindrical, and their sizes are adapted to the respective objects for which they are intended to secure. They have round or saucer heads, according to the purposes for which they may be intended; and the points are fore-locked or clinched on rings to prevent their drawing. Those for bolting the frames or beams together are generally square.

BOTTOM.—All that part of a ship or vessel that is below the Wales. Hence we use the epithet *sharp-bottomed* for vessels intended for quick-sailing, and *full-bottomed* for such as are designed to carry large cargoes.

BOW.—The circular part of the ship forward, terminated at the rabbet of the stem.

BRACES.—Straps of iron, copper, or mixed metal, secured with bolts and screws in the stern post and bottom planks. In their after ends are holes to receive the pintles by which the rudder is hung.

BREADTH.—A term more particularly applied to some essential dimensions of the extent of a ship or vessel athwartships, as the BREADTH-EXTREME, and the BREADTH-MOULDED, which are two of the principal dimensions given in the building of the ship. The *Extreme Breadth* is the extent of the midships or dead-flat with the thickness of the bottom plank included. The *Breadth-Moulded*, is the same extent without the thickness of the plank.

BREADTH-LINE.—A curved line of the ship lengthwise, intersecting the timbers at their respective broadest parts.

BREAK.—The sudden termination or rise in the decks of some merchant ships, where the aft and sometimes the fore part of the deck is kept up to give more height between decks, as likewise at the Drifts.

BREAST-HOOKS.—Large pieces of compass timber fixed within and athwart the bows of the ship, of which they are the principal security, and through which they are well bolted. There is generally one between each deck, and three or four below the lower deck, fayed upon the plank. Those below are placed square to the shape of the ship at their respective places. The Breast-Hooks that receive the ends of the deck-planks are also called DECK-HOOKS, and are fayed close home to the timbers in the direction of the decks.

BROKEN-BACKED OR HOGGED.—The condition of a ship when the sheer has departed from the regular and pleasing curve with which it was originally built. This is often occasioned by the improper situation of the centre of gravity, when so posited as not to counterbalance the effort of the water in sustaining the ship, or, by a great strain, or, from the weakness of construction. The latter is the most common circumstance, particularly in some Clipper ships, owing partly to their great length, sharpness of floor, or general want of strength in the junction of the component parts.

BUM-KIN, OR MORE PROPERLY BOOM-KIN.—A projecting piece of oak or fir, on each bow of a ship, fayed down upon the False Rail, or Rail of the Head, with its heel cleated against the Knight-head in large, and the bow in small, ships. It is secured outwards by an iron rod or rope lashing, which confines it downward to the knee or bow, and is used for the purpose of hauling down the fore-tack of the fore-sail.

BURTHEN.—The weight or measure that any ship will carry or contain when fit for sea.

BUTT.—The joints of the planks endwise, also the opening between the ends of the planks when worked for caulking. Where caulking is not used, the butts are rabbetted, and must fay close.

BUTTOCK.—That rounding part of the body abaft bounded by the fashion-pieces; and, at the upper part, by the wing transom.

BUTTOCK-LINES.—(On the Sheer Draught.) Curves, lengthwise, representing the ship as cut in vertical sections.

CAMBER.—Hollow or arching upwards. The decks are said to be *cambered* when their height increase toward the middle, from stem to stern, in the direction of the ship's length.

CAMEL.—A machine for lifting ships over a bank or shoal, originally invented by the celebrated De Witt, for the purpose of conveying large vessels from Amsterdam over the Pampus. They were introduced into Russia by Peter the Great, who obtained the model when he worked in Holland, as a common shipwright, and are now used at St. Petersburgh for lifting ships of war built there over the bar of the harbour. A Camel is composed of two separate parts, whose outsides are perpendicular, and insides concave, shaped so as to embrace the hull of a ship on both sides. Each part has a small cabin, with sixteen pumps and ten plugs, and contains twenty men. They are braced to the underpart of the ship by means of cables, and entirely inclose its sides and bottom. Being then towed to the bar, the plugs are opened, and the water admitted until the Camel sinks with the ship, and runs aground. Then, the water being pumped out, the Camel rises, lifts up the vessel, and the whole is towed over the bar. This machine can raise the ship eleven feet, or in other words, make it draw eleven feet less water.

CANT.—A term signifying the inclination that any thing has from a square or perpendicular. Hence the shipwrights say,

CANT-RIBBANDS, are those ribbands that do not lie in a horizontal or level direction, or square from the middle line, but nearly square from the timbers, as the diagonal ribbands. *See* RIBBANDS.

CANT-TIMBERS, are those timbers afore and abaft, whose planes are not square with, or perpendicular to, the middle-line of the ship.

CAPS.—Square pieces of oak, laid upon the upper blocks on which the ship is built, to receive the keel. They should be of the most freely grained oak, that they may be easily split out when the false keel is to be placed beneath. The depth of them may be a few inches more than the thickness of the false keel, that it may be set up close to the main keel by slices, &c.

A CAP SCUTTLE.—A framing composed of coamings and head ledges, raised above the deck, with a flat or top which shuts closely over into a rabbet.

CARLINGS.—Long pieces of timber, above four inches square, which lie fore and aft, in tiers, from beam to beam, into which their ends are scored. They receive the ends of the ledges for framing the decks. The Carlings by the side of, and for the support of, the mast, which receive the framing round the mast called the partners, are much larger than the rest, and are named the MAST CARLINGS. Besides these there are others, as the PUMP CARLINGS, which go next without the Mast Carlings, and between which the pumps pass into the well; and also the Fire-Hearth Carlings, that let up under the beam on which the Galley stands, with pillars underneath, and chocks upon it, fayed up to the ledges for support.

CARVEL WORK.—A term applied to Cutters and Boats, signifying that the seams of the bottom-planking are square, and to be tight by caulking as those of ships. It is opposed to the phrase CLINCHER-BUILT, which see.

CAULKING.—Forcing oakum into the seams and between the butts of the plank, &c., with iron instruments, in order to prevent the water penetrating into the ship.

CEILING OR FOOTWALING.—The inside planks of the bottom of the ship.

CENTRE OF CAVITY, OR OF DISPLACEMENT.—The centre of that part of the ship's body which is immersed in the water; and which is also the centre of the vertical force that the water exerts to support the vessel.

CENTRE OF GRAVITY.—That point about which all the parts of a body do, in any situation, exactly balance each other. Hence, 1. If a body be suspended by this point as the centre of motion, it will remain at rest in any position indifferently. 2. If a body be suspended in any other point, it can rest only in two positions, viz. when the centre of gravity is exactly above or below the point of suspension. 3. When the centre of gravity is supported, the whole body is kept from falling. 4. Because this point has a constant tendency to descend to the centre of the earth, therefore—5. When the point is at liberty to descend, the whole body must also descend, either by sliding, rolling, or tumbling over.

CENTRE OF MOTION.—That point of a body which remains at rest whilst all the other parts are in motion about it; and this is the same, in bodies of one uniform density throughout, as the centre of gravity.

CENTRE OF OSCILLATION.—That point in the axis or line of suspension of a vibrating body, or system of bodies, in which if the whole matter or weight be collected, the vibrations will still be performed in the same time, and with the same angular velocity, as before.

CENTRE OF PERCUSSION, in a moving body, is that point where the percussion or stroke is the greatest, and in which the whole percutient force of the body is supposed to be collected. PERCUSSION is the impression a body makes in falling or striking upon another, or the shock of bodies in motion striking against each other. It is either direct or oblique; *direct* when the impulse is given in a line perpendicular to the point of contact; and *oblique* when it is given in a line oblique to the point of contact.

CENTRE OF RESISTANCE TO A FLUID.—That point in a plane to which, if a contrary force be applied, it shall just sustain the resistance.

CHAIN OR CHAINS.—The links of iron which are connected to the binding that surround the dead-eyes of the channels. They are secured to the ship's side by a bolt through the toe-link called a *chain-bolt*.

CHAIN-BOLT.—A large bolt to secure the chains of the dead-eyes, for the purpose of securing the masts by the shrouds.

CHAIN-PLATES.—Thick iron plates, sometimes used, which are bolted to the ship's sides, instead of chains to the dead-eyes, as above.

CHAMFERING.—Taking off the sharp edge from timber or plank, or cutting the edge or end of any thing bevel or aslope.

CHANNELS.—The broad projection or assemblage of planks, fayed and bolted to the ship's sides, for the purpose of spreading the shrouds with a greater angle to the dead-eyes. They should therefore be placed either above or below the upper deck ports, as may be most convenient. But it is to be observed that, if placed too high, they strain the sides too much; and if placed too low, the shrouds cannot be made to clear the ports without difficulty. Their disposition will therefore depend on that particular which will produce the greatest advantage. They should fay to the sides only where the bolts come through, having an open space of about two inches in the rest of their length, to admit a free current of air, and a passage for wet and dirt, in order to prevent the sides from rotting.

CHANNEL WALES.—Three or four thick strakes, worked between the upper and lower deck ports in two decked ships, and between the upper and middle deck ports in three decked ships, for the purpose of strengthening the topside. They should be placed in the best manner for receiving the chain and preventer bolts, the fastenings of the deck-knees, &c.

CHEEKS.—Knees of oak-timber which support the knee of the head, and which they also ornament by their shape and mouldings. They form the basis of the head, and connect the whole to the bows, through which and the knee they are bolted.

*CHESTREES.—Pieces of oak timber fayed and bolted to the topsides, one on each side, abaft the fore-channels, with a sheave fitted in the upper part for the convenience of hauling home the main tack.

CHINE.—That part of the waterways, which is left the thickest, and above the deck-plank. It is bearded back that the lower seam of spirketting may be more conveniently caulked, and is gouged hollow in front to form a watercourse.

TO CHINSE.—To caulk slightly with a knife or chisel, those seams or openings that will not bear the force required for caulking in a more proper manner.

CLAMPS.—Those substantial strakes worked within side the ship, upon which the ends of the beams are placed.

CLEAN.—A term generally used to express the acuteness or sharpness of a ship's body: as when a ship is formed very acute or sharp forward, and the same aft, she is said to be *clean* both forward and aft.

CLINCHER-BUILT.—A term applied to the construction of some vessels and boats, when the planks of the bottom are so disposed, that the lower edge of every plank overlays the next under it, and the fastenings go through and clinch or turn upon the timbers. It is opposed to the term CARVEL-WORK.

CLINCHING OR CLENCHING.—Spreading the point of the bolt upon a ring, &c., by beating it with a hammer, in order to prevent its drawing.

COAMINGS.—The raised borders of oak about the edge of the hatches and scuttles, which prevent water from flowing down from off the deck. Their inside upper edge has a rabbet to receive the gratings.

COMPANION.—In ships of war, the framing and sash lights upon the Quarter-Deck or Poop, through which the light passes to the Commander's apartments. In merchant ships it is the birthing or hood round the ladder way, leading to the master's cabin, and in small ships is chiefly for the purpose of keeping the sea from beating down.

CONVERSION.—The art of lining and moulding timber, plank, &c., with the least possible waste.

COPING.—Turning the ends of iron lodging knees so as they may hook into the beams.

COUNTER.—A part of the Stern ; the *Lower Counter* being that arched part of the stern immediately above the wing transom. Above the Lower Counter is the *Second Counter*, the upper part of which is the under part of the Lights or Windows. The Counters are parted by their rails, as the lower counter springs from the tuck-rail, and is terminated on the upper part by the lower counter-rail. From the upper part of the latter springs the upper or second counter, its upper part terminating in the upper counter rail, which is immediately under the Lights.

COUNTER-SUNK.—The hollows in iron-plates, &c., which are excavated by an instrument called a Counter Sunk Bitt, to receive the heads of screws or nails so that they may be flush or even with the surface.

COUNTER TIMBERS.—The right-aft timbers which form the stern. The longest run up and form the lights, while the shorter only run up to the under part of them, and help to strengthen the counter. The side counter timbers are mostly formed of two pieces

---

* Obsolete.

scarphed together in consequence of their peculiar shape, as they not only form the right-aft figure of the stern, but partake of the shape of the topside also.

COVE.—The arch moulding sunk in at the foot or lower part of the taffarel.

CRAB.—A sort of little Capstan, formed of a kind of wooden pillar, whose lower end works in a socket, whilst the middle traverses or turns round in partners which clip it in a circle. In its upper end are two holes to receive bars, which act as levers, and by which it is turned round and serves as a capstan for raising of weights, &c. By a machine of this kind, so simple in its construction, may be hove up the frame timbers, &c., of vessels when building. For this purpose it is placed between two floor timbers, while the partners which clip it in the middle may be of four or five inch plank fastened on the same floors. A block is fastened beneath in the slip, with a central hole for its lower end to work in. Besides the Crab here described, there is another sort, which is shorter and portable. The latter is fitted in a frame composed of cheeks, across which are the partners, and at the bottom a little platform to receive the spindle.

CRADLE.—A strong frame of timber, &c., placed under the bottom of a ship in order to conduct her steadily in her ways till she is safely launched into water sufficient to float her.

CRANK.—A term applied to ships built too deep in proportion to their breadth, and from which they are in danger of over-setting.

CROAKY.—A term applied to plank when it curves or compasses much in short lengths.

CROSS SPALES.—Deals or fir plank nailed in a temporary manner to the frames of the ship at a certain height, and by which the frames are kept to their proper breadths, until the deck-knees are fastened. The main and top-timber breadths are the heights mostly taken for spalling the frames, but the height of the ports is much better, yet this may be thought too high if the ship is long in building.

CRUTCHES OR CLUTCHES.—The crooked timbers fayed and bolted upon the foot-waling abaft for the security of the heels of the half-timbers. Also stantions of iron or wood whose upper parts are forked to receive rails, spare masts, yards, &c.

CUP.—A solid piece of cast-iron, let into the step of the Capstan, and in which the iron spindle at the heel of the capstan works. *See Capstan.*

CUTTING-DOWN LINE.—The elliptical curve line, forming the upperside of the floor-timbers at the middle line. Also the line that forms the upper part of the Knee of the Head above the Cheeks. The cutting down line is represented as limiting the depth of every floor timber at the middle line, and also the height of the upper part of the dead-wood afore and abaft.

DAGGER.—A piece of timber that faces on to the poppets of the bilgeways, and crosses them diagonally, to keep them together. The plank that secures the heads of the poppets is called the *Dagger Plank*. The word *Dagger* seems to apply to any thing that stands diagonally or aslant.

DAGGER-KNEES.—Knees to supply the place of hanging knees. Their side arms are brought up aslant or nearly to the underside of the beams adjoining. They are chiefly used to the lower deck beams of merchant ships, in order to preserve as much stowage in the hold as possible. Any strait hanging knees not perpendicular to the side of the beam are in general termed *Dagger-Knees*.

DEAD-FLAT.—A name given to that timber or frame which has the greatest breadth and capacity in the ship, and which is generally called the *Midship Bend*. In those ships where there are several frames or timbers of equal breadth or capacity, that which is in the middle should be always considered as *Dead-Flat*, and distinguished as such by this character +. The timbers before the Dead-Flat are marked A, B, C, &c., in order; and those abaft Dead-Flat by the figures 1, 2, 3, &c. The Timbers adjacent to Dead-Flat, and of the same dimensions nearly, are distinguished by the characters (A) (B) &c. and (1) (2) &c.

DEAD-RISING, OR RISING LINE OF THE FLOOR.—Those parts of the floor or bottom, throughout the ship's length, where the sweep or curve at the head of the floor timber is terminated or inflects to join the keel. Hence, although the rising of the floor at the midship-flat is but a few inches above the keel at that place, its height forward and aft increases according to the sharpness of form in the body. Therefore the rising of the floor in the *sheer plan*, is a curved line drawn at the height of the ends of the floor timbers; and limited at the main frame, or dead-flat by the dead rising: appearing in flat ships nearly parallel to the keel for some timbers afore and abaft the midship frame; for which reason these timbers are called *flats:* but in sharp ships it rises gradually from the main frame, and ends on the stem and post.

DEAD-WATER.—The eddy water which the ship draws after her at her seat or line of floatation in the water, particularly close aft. To this particular, great attention should be paid in the construction of a vessel, especially in those with square tucks, for such being carried too low in the water, will be attended with great eddies or much *dead-water.*— Vessels with a round buttock have but little or no dead-water, because, by the rounding or arching of such vessels abaft, the water more easily recovers its state of rest.

DEAD-WOOD.—That part of the basis of a ship's body, forward and aft, which is formed by solid pieces of timber scarfed together lengthwise on the keel. These should be sufficiently broad to admit of a stepping or rabbet for the heels of the timbers, that the latter may not be continued downwards to sharp edges; and they should be sufficiently high to seat the floors. Afore and abaft the floors the dead-wood is continued to the cutting down line, for the purpose of securing the heels of the Cant-timbers.

DEPTH IN THE HOLD.—The height between the floor and the lower deck. This is one of the principal dimensions given for the construction of a ship. It varies according to the height at which the guns are required to be carried from the water; or, according to the trade for which a vessel is designed.

DIAGONAL LINE.—A line cutting the body-plan diagonally from the timbers to the middle line. It is square with, or perpendicular to, the shape of the timbers, or nearly so, till it meets the Middle Line.

DIAGONAL RIBBAND.—A narrow plank, made to a line formed on the Half-breadth-plan, by taking the intersections of the diagonal line with the timbers in the body-plan to where it cuts the middle line in its direction, and applying it to their respective stations on the Half-breadth-plan, which forms a curve to which the ribband is made as far as the Cant Body extends, and the square frame adjoining.

DOG.—An iron implement used by shipwrights, having a fang at one, or sometimes at each, end, to be driven into any piece for supporting it while hewing, &c. Another sort has a fang in one end and an eye in the other, in which a rope may be fastened, and used to haul any thing along.

DOG SHORE.—A shore particularly used in Launching.

DOUBLING.—Planking of ships' bottoms twice. It is sometimes done to new ships when the original planking is thought to be too thin; and, in repairs, it strengthens the ship, without driving out the former fastenings.

DRAUGHT.—The drawing or design of the ship, upon paper, describing the different parts, and from which the ship is to be built. It is mostly drawn by a scale of one quarter of an inch to a foot, so divided or graduated that the dimensions may be taken to one inch.

DRAUGHT OF WATER.—The Depth of water a ship displaces when she is afloat.

DRIVER.—The foremost spur on the bilgeways; the heel of which is fayed to the foreside of the foremost poppet, and cleated on the bilgeways, and the sides of it stand fore and aft. It is now seldom used.

DRUMHEAD.—The head of a capstan, formed of semi-circular pieces of elm, which, framed together, form the circle into which the capstan-bars are fixed.

DRUXEY.—A state of decay in timber with white spungy veins, the most deceptive of any defect.

EDGING OF PLANK.—Sawing or hewing it narrower.

EKEING.—Making good a deficiency in the length of any piece by scarphing or butting, as at the end of deck-hooks, cheeks, or knees. The *Ekeing* at the lower part of the Supporter under the Cathead, is only to continue the shape and fashion of that part, being of no other service. We make this remark because, if the Supporter were stopt short without an ekeing, it would be the better as it causes the side to rot, and it commonly appears fair to the eye in but one direction. The EKEING is also the piece of carved work under the lower part of the Quarter-piece at the aft part of the Quarter-gallery.

ELEVATION.—The orthographic draught, or perpendicular plan of a ship, whereon the heights and lengths are expressed. It is called by shipwrights the SHEER-DRAUGHT.

ENTRANCE.—A term applied to the fore part of the ship under the load-water line; as, "She has a fine entrance," &c.

EVEN KEEL.—A ship is said to swim on an even keel when she draws the same quantity of water abaft as forwards.

FACING.—Letting one piece, about an inch in thickness, on to another, in order to strengthen it.

FAIR.—A term to denote the evenness or regularity of a curve or line.

FALLING-HOME, or, by some, TUMBLING HOME.—The inclination which the topside has within from a perpendicular.

FALSE-KEEL.—A second keel, composed of elm-plank, or thick stuff, fastened in a slight manner under the main keel, to prevent it from being rubbed. Its advantages also are, that, if the ship should strike the ground, the false keel will give way, and thus the main keel will be saved; and it will be the means of causing the ship to hold the wind better.

FALSE-POST.—A piece tabled on to the after part of the heel of the main part of the stern post. It is to assist the conversion and preserve the main post should the ship tail aground.

FALSE-RAIL.—A rail fayed down upon the upper side of the main, or upper rail of the head. It is to strengthen the head-rail, and forms the seat of ease at the after end next the bow.

FASHION PIECES.—The timbers so called from their fashioning the after part of the ship in the plane of projection, by terminating the breadth and forming the shape of the stern. They are united to the ends of the transoms and to the dead-wood.

TO FAY.—To join one piece so close to another that there shall be no perceptible space between them.

FILLING-TIMBERS.—The intermediate timbers between the frames that are gotten up into their places singly after the frames are ribbanded and shored.

FLAIRING.—The reverse of *Falling* or *Tumbling Home*. As this can be only in the fore-part of the ship, it is said that a ship has a *flairing-bow*, when the topside falls outward from a perpendicular. Its uses are, to shorten the Cathead, and yet keep the anchor clear of the bow. It also prevents the sea from breaking in upon the Forecastle.

FLATS.—A name given to the timbers a-midships that have no bevellings, and are similar to dead-flat, which is distinguished by this character ×. *See* Dead-flat.

FLOOR.—The bottom of a ship, or all that part on each side of the keel which approaches nearer to a horizontal than a perpendicular direction, and whereon the ship rests when aground.

FLOORS, OR FLOOR TIMBERS.—The timbers that are fixed athwart the keel, and upon which the whole frame is erected. They generally extend as far forward as the fore-mast, and as far aft as the after square timber; and sometimes, one or two cant-floors are added.

FLUSH.—With a continued even surface: As a flush deck, which is a deck upon one continued line, without interruption, from fore to aft.

FORE BODY.—That part of the ship's body, afore the Midships or Dead-flat. *See* Bodies. This term is more particularly used in expressing the *figure* or *shape* of that part of the ship.

FORE-FOOT.—The foremost piece of the keel.

FORE-LOCK.—A thin circular wedge of iron, used to retain a bolt in its place, by being thrust through a mortise hole at the point of the bolt. It is sometimes turned or twisted round the bolt to prevent its drawing.

FORE PEEK.—Close forward under the lower deck.

FRAMES.—The bends of timber which form the body of the ship; each of which is composed of one *floor-timber*, two or three *futtocks*, and a *top-timber* on each side; which, being united together, form the frame. Of these frames, or bends, that which encloses the greatest space is called the *midship* or *main frame* or *bend*. The arms of the floor timber form a very obtuse angle; and in the other frames, this angle decreases or gradually becomes sharper, fore and aft, with the middle line of the ship. Those floors which form the acute angles afore and abaft are called the *Rising Floors*. A frame of timbers is commonly formed by arches of circles called *Sweeps*, of which there are generally five. 1st. The *Floor Sweep*, which is limited by a line in the Body Plan perpendicular to the plane of elevation, a little above the keel; and the height of this line above the keel is called the *Dead Rising*. The upper part of this arch forms the head of the floor timber. 2nd. The *Lower Breadth Sweep*; the centre of which is in the line representing the lower height of breadth. 3rd. The *Reconciling Sweep*; this sweep joins the two former, without intersecting either; and makes a fair curve from the lower height of breadth to the rising line. If a straight line be drawn from the upper edge of the keel to touch the back

B

of the floor sweep, the form of the midship frame below the lower height of breadth will be obtained. 4th. The *Upper Breadth Sweep;* the centre of which is in the line representing the upper height of breadth of the timber. This sweep described upwards forms the lower part of the top-timber. 5th. The *Top-Timber Sweep,* or *Back Sweep*, is that which forms the hollow of the top-timber. This hollow is, however, very often formed by a mould, so placed as to touch the upper breadth sweep, and pass through the point limiting the half-breadth of the top-timber.

FRAME TIMBERS.—The various timbers that compose a frame bend; as the floor timber, the first, second, third, and fourth, futtocks, and top timber, which are united, by a proper shift, to each other, and bolted through each shift. They are often kept open, for the advantage of the air, and fillings fayed between them in wake of the bolts. Some ships are composed of frames only, and are supposed to be of equal strength with others of larger scantling.

FUTTOCKS.—The separate pieces of timber of which the frame timbers are composed. They are named according to their situation, that nearest the keel being called the first futtock, the next above, the second futtock, &c.

GARBOARD STRAKE.—That strake of the bottom which is wrought next the keel, and rabbets therein.

GRIPE.—A piece of elm timber that completes the lower part of the knee of the head, and makes a finish with the fore-foot. It bolts to the stem, and is farther secured by two plates of copper in the form of a horse-shoe, and therefrom called by that name.

GROUNDWAYS.—Large pieces of timber, generally defective, which are laid upon piles driven in the ground, across the dock or slip, in order to make a good foundation to lay the blocks on, upon which the ship is to rest.

GUNWALE.—That horizontal plank which covers the heads of the timbers between the main and fore drifts.

HALF-TIMBERS.—The short timbers in the cant bodies which are answerable to the lower futtocks in the square body.

HANGING-KNEE.—Those knees against the sides whose arms hang vertically or perpendicular.

HARPINS.—Pieces of oak, similar to ribbands, but trimmed and bevelled to the shape of the body of the ship, and holding the fore and after cant bodies together until the ship is planked. But this term is mostly applicable to those at the bow; hence arises the phrase "lean and full harpin," as the ship at this part is more or less acute.

HEAD.—The upper end of any thing; but more particularly applied to all the work fitted afore the stem, as the Figure, the Knee, Rails, &c. A "Scroll Head" signifies that there is no carved or ornamental figure at the head, but that the termination is formed and finished off by a *valute*, or scroll turning outwards. A "Fiddle Head" signifies a similar kind of finish, but with the scroll turning aft or inwards.

HEAD-LEDGES.—The 'thwartship pieces which frame the hatchways and ladderways.

HEAD-RAILS.—Those rails in the Head which extend from the back of the figure to the cathead and bows, which are not only ornamental to the frame, but useful to that part of the ship.

HEEL.—The lower end of a tree, timber, &c. A ship is also said to *Heel* when she is not upright but inclines under a side pressure.

HOGGING.—*See* BROKEN BACKED. A ship is said to *Hog* when the middle part of her keel and bottom are so strained as to curve or arch upwards. This term is therefore opposed to *Sagging*, which, applied in a similar manner, means, by a different sort of strain, to curve downwards.

HOLD.—That part of the ship below the lower deck, between the bulkheads, which is reserved for the stowage of ballast, water, and provisions, in ships of war, and for that of the cargo in merchant-vessels.

HOODING ENDS.—Those ends of the planks which bury in the rabbets of the stem and stern post.

HORSE-IRON.—An iron fixed in a handle, and used with a beetle by caulkers, to *horse-up* or harden in the oakum.

HORSE-SHOES.—Large straps of iron or copper shaped like a horse-shoe and let into the stem and gripe on opposite sides, through which they are bolted together to secure the gripe to the stem.

HULL.—The whole frame or body of a ship, exclusive of the masts, yards, sails, and rigging.

IN AND OUT.—A term sometimes used for the scantling of the Timbers the moulding way, but more particularly applied to those bolts in the knees, riders, &c., which are driven through the ship's sides, or athwartships, and therefore called *"In and out bolts."*

INNER POST.—A piece of oak timber, brought on and fayed to the foreside of the main stern-post, for the purpose of seating the Transoms upon it. It is a great security to the ends of the planks, as the main post is seldom sufficiently afore the rabbet for that purpose, and is also a great strengthener to that part of the ship.

KEEL.—The main and lowest timber of a ship, extending longitudinally from the stem to the stern post. It is formed of several pieces, which are scarphed together endways, and form the basis of the whole structure. Of course it is usually the first thing laid down upon the blocks for the construction of the ship.

KEELSON or, MORE COMMONLY, KELSON.—The timber, formed of long square pieces of oak, fixed within the ship exactly over the keel, (and which may therefore be considered as the counter part of the latter) for binding and strengthening the lower part of the ship; for which purpose it is fitted to, and laid upon, the middle of the floor timbers, and bolted through the floors and keel.

KNEES. The crooked pieces of oak timber by which the ends of the beams are secured to the sides of the ship. Of these, such as are fayed vertically to the sides are called *Hanging-Knees*, and such as are fixed parallel to, or with the hang of, the deck, are called *Lodging-Knees*.

KNEE OF THE HEAD.—The large flat timber fayed edgeways upon the fore-part of the stem. It is formed by an assemblage of pieces of oak coaked or tabled together edgewise, by reason of its breadth, and it projects the length of the Head. Its fore part should form a handsome serpentine line, or inflected curve. The principal pieces are named the *Main-piece* and *Lacing*.

LABOURSOME.—Subject to *labour*, or to pitch and roll violently in a heavy sea, by which the masts and even the hull may be endangered. For, by a series of heavy rolls the rigging becomes loosened, and the masts at the same time may strain upon the shrouds with an effort which they will be unable to resist; to which may be added, that the continual agitation of the vessel loosens her joints, and makes her extremely leaky.

TO LAP OVER or UPON.—The mast carlings are said to lap upon the beams by reason of their great depth, and head-ledges at the ends lap over the coamings.

LAP-SIDED.—A term expressive of the condition of a vessel when she will not swim upright, owing to her sides being unequal.

LAUNCHING-PLANKS.—A set of planks mostly used to form the platform on each side of the ship, whereon the bilgeways slide for the purpose of launching.

LAYING-OFF, OR LAYING-DOWN.—The act of delineating the various parts of the ship, to its true size, upon the mould-loft floor, from the draught given for the purpose of making the moulds.

LEDGES.—Oak or fir scantling used in framing the decks, which are let into the carlings athwartships. The ledges for gratings are similar, but arch or round up agreeably to the head ledges.

LENGTHENING.—The operation of separating a ship athwartships and adding a certain portion to her length. It is performed by clearing or driving out all the fastenings in wake of the butts of those planks which may be retained, and the others are cut through. The after end is then drawn apart to a limited distance equal to the additional length proposed. The Keel is then made good, the floors crossed, and a sufficient number of timbers raised to fill up the vacancy produced by the separation. The kelson is then replaced to give good shift to the new scarphs of the Keel, and as many beams as may be necessary are placed across the ship in the new interval, and the planks on the outside are replaced with a proper shift. The clamps and footwaling within the ship are then supplied, the beams knee'd, and the ship completed in all respects as before.

TO LET-IN.—To fix or fit one timber or plank into another, as the ends of carlings into the beams, and the beams into the clamps, vacancies being made in each to receive the other.

LEVEL LINES.—Lines determining the shape of a ship's body horizontally, or square from the middle line of the ship.

LIMBER PASSAGE.—A passage or channel formed throughout the whole length of the floor, on each side of the kelson, for giving water a free communication to the pumps. It is formed by the LIMBER-STRAKE on each side, a thick strake wrought next the kelson, from the upper side of which the depth in the hold is always taken. This strake is kept at about eleven inches from the kelson, and forms the passage fore and aft which admits the water with a fair run to the pump-well. The upper part of the Limber Passage is formed by the LIMBER-BOARDS, which are made to keep out all dirt and other obstructions. These boards are composed of short pieces of oak plank, one edge of which is fitted by a rabbet into the limber strake, and the other edge bevelled with a descent against the kelson. They are fitted in short pieces for the convenience of taking up any one, or more, readily, in order to clear away any obstruction in the passage. When the limber boards are fitted, care should be taken to have the butts in those places where the bulkheads come, as there will be then no difficulty in taking those up which come near the bulkheads. A hole is bored in the middle of each butt to admit the end of a crow for prizing it up when required. To prevent the boards from being displaced, each should be marked with a line corresponding with one on the Limber Strake.

LIMBER HOLES are square grooves cut through the underside of the floor timber, about nine inches from the side of the Keel on each side, through which water may run toward the pumps, in the whole length of the floors. This precaution is requisite in merchant ships only, where small quantities of water, by the heeling of the ship, may come through the ceiling and damage the cargo. It is for this reason that the lower futtocks of merchant ships are cut off short of the Keel.

LIPS OF SCARPHS.—The substance left at the ends, which would otherwise become sharp, and be liable to split; and, in other cases could not bear caulking as the scarphs of the keel, stem, &c.

MAIN BREADTH.—The broadest part of the ship at any particular timber or frame, which is distinguished on the sheer-draught by the upper and lower heights of breadth lines.

MAIN WALES.—The lower Wales; which are generally placed on the lower breadth, and so that the main-deck knee-bolts may come into them.

MANGER.—An apartment extending athwart the ship immediately within the hawse-holes. It serves as a fence to interrupt the passage of water which may come in at the hawse-holes, or from the cable when heaving in; and the water thus prevented from running aft is returned into the sea by the manger scuppers, which are larger than the other scuppers on that account.

MAULS.—Large hammers used for driving treenails, having a steel face at one end, and a point or pen drawn out at the other. Double-headed Mauls have a steel face at each end, of the same size, and are used for driving of bolts, &c.

META-CENTRE.—That point in a ship above which the centre of gravity must by no means be placed ; because, if it were, the vessel would be liable to overset. The *meta-centre*, which has also been called the *shifting-centre*, depends upon the situation of the centre of cavity ; for it is that point where a vertical line drawn from the centre of cavity cuts a line passing through the centre of gravity, and being perpendicular to the Keel.

MIDDLE LINE.—A line dividing the ship exactly in the middle. In the horizontal or half-breadth plan it is a right line bisecting the ship from the stem to the stern-post; and, in the plane of projection, or body plan, it is a perpendicular line bisecting the ship from the keel to the height of the top of the side.

MOMENTA, or MOMENTS.—The plural of *Momentum*.

MOMENTUM of a heavy body, or of any extent considered as a heavy body, is the product of the weight multiplied by the distance of its centre of gravity from a certain point, assumed at pleasure, which is called the centre of the momentum, or from a line which is called the axis of the momentum.

MORTISE.—A hole or hollow made of a certain size and depth in a piece of timber, &c. in order to receive the end of another piece with a tenon fitted exactly to fill it.

MOULDS.—Pieces of deal or board made to the shape of the lines on the Mould Loft Floor, as the Timbers, Harpins, Ribbands, &c. for the purpose of cutting out the different pieces of timber, &c., for the ship. Also the thin flexible pieces of pear-tree or box used in constructing the draughts and plans of ship, which are made in various shapes ; viz. to the segments of circles from one foot to 22 feet radius, increasing six inches on each edge, and numerous elliptical curves, with other figures.

MOULDED.—Cut to the mould. Also the size or bigness of the timbers that way the mould is laid. *See* SIDED.

NAILS.—Iron pins of various descriptions for fastening board, plank, or iron work; viz: *Deck Nails*, or *Spike Nails*, which are from 4 inches and a half to 12 inches long, have snug heads, and are used for fastening planks and the flat of the decks. *Weight Nails* are similar to deck nails, but not so fine, have square heads, and are used for fastening cleats, &c. *Ribband Nails* are similar to weight nails, with this difference, that they have large round heads, so as to be more easily drawn. They are used for fastening the ribbands, &c. *Clamp Nails* are short stout nails, with large heads, for fastening iron clamps. *Port Nails*, double and single, are similar to clamp nails, and used for fastening iron work.—*Rudder Nails* are also similar, but used chiefly for fastening the pintles and braces. *Filling Nails* are generally of cast iron, and driven very thick in the bottom planks instead of copper sheathing.* *Sheathing Nails* are used to fasten wood sheathing on the ship's bottom, to preserve the plank, and prevent the filling nails from tearing it too much.* *Nails of sorts* are 4, 6, 8, 10, 24, 30, and 40 penny nails, all of different lengths, and used for nailing board, &c. *Scupper Nails* are short nails, with very broad heads, used to nail the flaps of the scuppers. *Lead Nails* are small round-headed nails for nailing of lead. *Flat Nails* are small sharp-pointed nails, with flat thin heads, for nailing the scarphs of moulds. *Sheathing Nails* for nailing copper sheathing are of metal, cast in moulds, about one inch and a quarter long; the heads are flat on the upper side and counter-sunk below: the upper side is polished to obviate the adhesion of weeds. *Boat Nails*, used by Boat-builders, are of various lengths, generally rose headed, square at the points, and made both of copper and iron.

OAKUM.—Old Rope, untwisted and loosened like hemp, in order to be used in caulking.

To OVER-LAUNCH.—To run the butt of one plank to a certain distance beyond the next butt above or beneath it, in order to make stronger work.

PALLETTING.—A slight platform, made above the bottom of the Magazine, to keep the powder from moisture.

PALLS.—Stout pieces of iron, so placed near a capstan or windlass as to prevent a recoil which would overpower the men at the bars when heaving.

PARTNERS.—Those pieces of thin plank, &c., fitted into a rabbet in the Mast or Capstan carlings for the purpose of wedging the mast and steadying the Capstan. Also any plank that is thick, or above the rest of the deck, for the purpose of steadying whatever passes through the deck, as the pumps, bowsprit, &c.

To PAY.—To lay on a coat of tar, &c., with a mop or brush, in order to preserve the wood and keep out water. When one or more pieces are scarphed together, as the beams, &c., the inside of the scarphs are paid with tar as a preservative; and the seams after they are caulked are payed with pitch to keep the water from the oakum, &c.

PINK.—A ship with a very narrow round stern; whence all vessels, however small, having their sterns fashioned in this manner, are said to be *pink-sterned*.

PINTLES.—Straps of mixed metal, or of iron, fastened on the rudder, in the same manner as the braces on the stern post, having a stout pin or hook at the ends, with the points downwards to enter in and rest upon the braces on which the rudder traverses or turns, as upon hinges, from side to side. Sometimes one or two are shorter than the rest, and work in a socket brace, whereby the rudder turns easier. The latter are called *Dumb-Pintles*. Some are bushed.

PITCH.—Tar, boiled to a harder and more tenacious substance.

PITCHING.—The inclination or vibration of the ship lengthwise about her centre of gravity; or the motion by which she plunges her head and after-part alternately into the hollow of the sea. This is a very dangerous motion, and when considerable, not only retards the ship's way, but endangers the masts and strains the vessel.

PLANKING.—Covering the outside of the timbers with plank; sometimes quaintly called *Skinning*, the plank being the outer coating, when the vessel is not sheathed.

PLANK-SHEERS, OR PLANK-SHEER.—The pieces of plank laid horizontally over the timber-heads of the Quarter-Deck and Forecastle, for the purpose of covering the top of the side, hence sometimes called Covering-Boards.

POINT-VELIQUE.—That point where, in a direct course, the centre of effort of all the sails should be found.

POPPETS.—Those pieces (mostly fir) which are fixed perpendicularly between the ship's bottom and the bilgeways, at the fore and aftermost parts of the ship, to support her in launching.

PUMP.—The machine, fitted in the wells of ships, to draw water out of the hold.

---
* Obsolete.

PUMP CISTERNS.—Cisterns fixed over the heads of the pumps, to receive the water until it is conveyed through the sides of the ship by the Pump-dales.

PUMP-DALES.—Pipes fitted to the cisterns, to convey the water from them through the ship's sides.

QUARTER-GALLERIES.—The projections from the Quarters abaft, fitted with sashes and ballusters, and intended both for convenience and ornament to the aft part of the ship.

To QUICKEN.—To give any thing a greater curve. For instance, "*To Quicken the Sheer*," is to shorten the radius by which the curve is struck. This term is therefore opposed to straightening the sheer.

RABBET.—A joint made by a groove, or channel, in a piece of timber cut for the purpose of receiving and securing the edge or ends of the planks, as the planks of the bottom into the keel, stem, or stern post, or the edge of one plank into another.

RAG-BOLT.—A sort of bolt having its point jagged or barbed to make it hold the more securely.

RAKE.—The overhanging of the stem or stern beyond a perpendicular with the keel, or any part or thing that forms an obtuse angle with the horizon.

RAM-LINE.—A small rope or line sometimes used for the purpose of forming the sheer or hang of the decks, for setting the beams, fair, &c.

RASING.—The act of marking by a mould on a piece of timber; or any marks made by a tool called a *rasing-knife*.

To RECONCILE.—To make one piece of work answer fair with the moulding or shape of the adjoining piece; and, more particularly, in the reversion of curves.

REEMING.—A term used by caulkers for opening the seams of the planks, that the oakum may be more readily admitted.

REEMING-IRONS.—The large irons used by caulkers in opening the seams.

RENDS.—Large open splits or shakes in timber; particularly in plank, occasioned by its being exposed to the wind or sun, &c.

RIBBANDS.—The longitudinal pieces of fir, about five inches square, nailed to the timbers of the square body (those of the same description in the Cant Body being shaped by a mould and called *Harpins*) to keep the body of the ship together, and in its proper shape, until the plank is brought on. The shores are placed beneath them. They are removed entirely when the planking comes on. The difference between *Cant Ribbands* and *Square* or *Horizontal Ribbands* is, that the latter are only ideal, and used in laying-off.

RIBBAND-LINES.—The same with diagonal lines.

RISING.—A term derived from the shape of a ship's bottom in general, which gradually narrows or becomes sharper towards the stem and the stern-post. On this account it is that the Floor, towards the extremities of the ship, is raised or lifted above the keel: otherwise the shape would be so very acute, as not to be provided from timber with sufficient strength in the middle, or cutting-down. The floor timbers forward and abaft, with regard to their general form and arrangement, are therefore gradually lifted or raised upon a solid body of wood called the *dead* or *rising wood*, which must, of course, have more or less rising as the body of the ship assumes more or less fullness or capacity. See DEAD RISING.

The RISING OF BOATS is a narrow strake of board fastened within side to support the thwarts.

RISING FLOORS.—The floors forward and abaft, which, on account of the rising of the body, are the most difficult to be obtained, as they must be deeper in the throat or at the cutting down to preserve strength.

RISING-LINE.—An elliptical line, drawn on the plan of elevation, to determine the sweep of the floor-heads throughout the ship's length, which accordingly ascertains the shape of the bottom with regard to its being full or sharp.

ROLLING.—That motion by which a ship vibrates from side to side. Rolling is therefore a sort of revolution about an imaginary axis passing through the centre of gravity of the ship: so that the nearer the centre of gravity is to the keel, the more violent will be the roll; because the centre about which the vibrations are made is placed so low in the bottom, that the resistance made by the keel to the volume of water which it displaces in rolling, bears very little proportion to the force of the vibration above the centre of gravity, the radius of which extends as high as the mast-heads   But, if the centre of gravity is placed higher above the keel, the radius of the vibration will not only be diminished, but such an additional force to oppose the motion of rolling will be communicated to that part of the ship's bottom as may contribute to diminish this movement considerably. It may be observed that, with respect to the formation of a ship's body, that shape which approaches nearest to a circle is the most liable to roll; as it is evident, that if this be agitated in the water, it will have nothing to restrain it; because the rolling or rotation about its centre displaces no more water than when it remains upright; and, hence, it becomes necessary to increase the depth of the keel, the rising of the floors, and the deadwood afore and abaft.

ROOM AND SPACE.—The distance from the moulding edge of one timber to the moulding edge of the next timber, which is always equal to the breadth of two timbers, and two to four inches more. The room and space of all ships that have ports should be so disposed, that the scantling of the timber on each side of the lower ports, and the size of the ports fore and aft, may be equal to the distance of two rooms and spaces.

ROUGH-TREE RAILS.—In men-of-war the broad plank running fore and aft covering the heads of the top timbers, thus forming the bottom of the hammock netting. In merchant vessels the rails along the waist and quarters, nearly breast high, to prevent persons from falling overboard. This term originated from the practice in merchant vessels of carrying their rough or spare gear in crutch irons along their waist.

RUDDER-CHOCKS.—Large pieces of fir, to fay or fill up the excavation on the side of the rudder in the rudder hole; so that the helm being in midships the rudder may be fixed, and supposing the tiller broke another might thus be replaced.

RUN.—The narrowing of the ship abaft, as of the floor towards the stern-post, where it becomes no broader than the post itself. This term is also used to signify the running or drawing of a line on the ship, or mould loft floor, as "to *run* the wale line," or deck line, &c.

SCANTLING.—The dimensions given for the timbers, planks, &c. Likewise all quartering under five inches square, which is termed Scantling; all above that size is called *Carling*.

SCARPHING.—The letting of one piece of timber or plank into another with a lap, in such a manner, that both may appear as one solid and even surface, as keel-pieces, stem-pieces, clamps, &c.

SCUPPERS.—Lead pipes let through the ship's side to convey the water from the decks.

SEAMS.—The openings between the edges of the planks when wrought.

SEASONING.—A term applied to a ship kept standing a certain time after she is completely framed and dubbed out for planking, which should never be less than six months when circumstances will permit. *Seasoned Plank or Timber* is such as has been cut down and sawn out one season at least, particularly when throughly dry, and not liable to shrink.

SEATING.—That part of the floor which fays on the dead-wood; and of a transom which fays against the post.

SENDING OR 'SCENDING.—The act of pitching violently into the hollows or intervals of the waves.

SETTING OR SETTING-TO.—The act of making the planks, &c., fay close to the timbers, by driving wedges between the plank, &c., and a wrain-staff. Hence we say, "set or set away," meaning to exert more strength. The power or engine used for the purpose of setting is called a SETT, and is composed of two ring-bolts, and a wrain-staff, cleats, and lashings.

SHAKEN OR SHAKY.—A natural defect in plank or timber when it is full of splits or clefts and will not bear fastening or caulking.

SHEATHING.—A thin sort of doubling, or casing, of fir-board or sheet copper, and sometimes of both, over the ship's bottom, to protect the planks from worms, &c. Tar and hair, or brown paper dipt in tar and oil, is laid between the sheathing and the bottom.

SHEER.—The longitudinal curve or hanging of ship's side in a fore and aft direction.

SHEER-DRAUGHT.—The plan of elevation of a ship, whereon is described the outboard works, as the wales. sheer-rails, ports, drifts, head, quarters, post, and stem, &c., the hang of each deck inside, the height of the water lines, &c.

SHEER-STRAKE.—The strake or strakes wrought in the topside, of which the upper edge is wrought well with the toptimber line, or top of the side, and the lower edge kept well with the upper part of the upper deck ports in midships, so as to be continued whole all fore and aft, and not cut by the ports. It forms the chief strength of the upper part of the topside, and is therefore always worked thicker than the other strakes, and scarphed with Hook and Butt between the drifts.

SIDING OR SIDED.—The size or dimensions of timber the contrary way to the moulding, or moulded side.

SIRMARKS.—The different places marked upon the moulds where the respective bevellings are to be applied, as the lower sirmark, floor sirmark, &c.

SLIDING PLANKS are the planks upon which the Bilgeways slide in Launching.

SLIP.—The foundation laid for the purpose of building the ship upon, and launching her.

To SNAPE.—To hance or bevel the end of any thing so as to fay upon an inclined plane.

SNYING.—A term applied to planks when their edges round or curve upwards. The great sny occasioned in full bows, or buttocks is only to be prevented by introducing Steelers.

SPECIFIC GRAVITY.—The comparative difference in the weight or gravity of two bodies of equal bulk; hence called also, relative or comparative gravity, because we judge of it by comparing one body with another.

A TABLE OF SPECIFIC GRAVITIES.

Lead 11325	Ebony 1177	Rain Water 1000
Fine Copper 9000	Pitch 1150	Oak 925
Gun Metal 8784	Rosin 1100	Ash 800
Fine Brass 8350	Mahogany 1063	Beech 700
Iron from....7827 to 7645	Box Wood 1030	Elm 600
Cast Iron 7425	Sea Water 1030	Fir 548
Sand 1520	Tar 1015	Cork 240
Lignum Vitæ 1327	River Water........ 1009	Common Air........ 1.232

These numbers being the weight of a cubic foot, or 1728 cubic inches, of each of the bodies in avoirdupois ounces; by proportion, the quantity in any other weight, or the weight of any other quantity, may be readily known.

For Example.—Required the content of an irregular piece of oak, which weighs 76lbs. or 1216 ounces.

$$\text{Sp. gr. oz. wt. oz. cub in. cub. in.}$$
Here as 925 : 1216 :: 1728 : 2271=1 ft. 543 inches cubic, the contents.

SPIRKETTING.—A thick strake, or strakes, wrought within side upon the ends of the beams or waterways. In ships that have ports the spirketting reaches from the waterways to the upper side of the lower sill, which is generally of two strakes, wrought anchor-stock fashion; in this case, the planks should always be such as will work as broad as possible, admitting the butts about six inches broad.

SQUARE BODY.—The figure which comprehends all the timbers whose areas or planes are perpendicular to the keel, which is all that portion of a ship between the cant-bodies. *See* BODIES.

SQUARE-TIMBERS.—The timbers which stand square with, or perpendicular to, the keel.

SQUARE TUCK.—A name given to the after part of a ship's bottom when terminated in the same direction up and down as the wing transom, and the planks of the bottom end in a rabbet at the foreside of the fashion-piece; whereas ships with a buttock are round or circular, and the planks of the bottom end upon the wing-transom.

STABILITY.—That quality which enables a ship to keep herself steadily in the water, without rolling or pitching. Stability, in the construction of a ship is only to be acquired, by fixing the centre of gravity at a certain distance below the meta-centre; because the stability of the vessel increases with the altitude of the meta-centre above the centre of gravity. But, when the meta-centre coincides with the centre of gravity, the vessel has no tendency whatever to remove out of the situation into which it may be put. Thus, if the vessel be inclined either to the starboard or larboard side, it will remain in that position till a new force is impressed upon it: in this case, therefore, the vessel would not be able to carry sail, and is consequently unfit for the purposes of navigation. If the meta-centre falls below the common centre of gravity, the vessel will immediately overset.

STEELER.—A name given to the foremost or aftermost plank, in a strake which drops short of the stem and stern post, and of which the end or butt nearest the rabbet is worked very narrow and well forward or aft. Their use is, to take out the snying edge occasioned by a full bow, or sudden circular buttock.

STEM.—The main timber at the fore part of the ship, formed, by the combination of several pieces, into a circular shape, and erected vertically to receive the ends of the bow-planks, which are united to it by means of a rabbet. Its lower end scarphs or boxes into the keel, through which the rabbet is also carried, and the bottom unites in the same manner.

STEMSON.—A piece of compass timber, wrought on the aft part of the apron within-side, the lower end of which scarphs into the kelson. Its upper end is continued as high as the middle or upper deck; and its use is to succour the scarphs of the apron, as that does those of the stem.

STEPS OF THE MASTS.—The steps into which the heels of the masts are fixed, are large pieces of timber. Those for the main and fore masts are fixed across the kelson, and that for the mizen mast upon the lower deck beams. The holes or mortises into which the masts step, should have sufficient wood on each side, to accord in strength with the tenon left at the heel of the mast, and the hole should be cut rather less than the tenon, as an allowance for shrinking.

STEPS FOR THE SHIP'S SIDE.—The pieces of quartering, with mouldings, nailed to the sides amidship, about nine inches asunder, from the Wale upwards, for the convenience of persons getting on board.

STERN FRAME.—The strong frame of timber, composed of the stern-post, transoms, and fashion-pieces, which form the basis of the whole stern.

STERN-POST.—The principal piece of timber in the stern frame on which the rudder is hung, and to which the transoms are bolted. It therefore terminates the ship below the wing-transom, and its lower end is tenoned into the keel.

STIVING.—The elevation of a ship's cathead or bowsprit; or the angle which either makes with the horizon, generally called Steeve.

STOPPINGS-UP.—The poppets, timber, &c. used to fill up the vacancy between the upper side of the bilgeways and the ship's bottom, for supporting her when launching.

STRAIGHT OF BREADTH.—The space before and abaft dead-flat, in which the ship is of the same uniform breadth, or of the same breadth as at ✕ or dead-flat. *See* DEAD-FLAT.

STRAKE.—One breadth of plank wrought from one end of the ship to the other, either within or without board.

TABLING.—Letting one piece of timber into another by alternate scores or projections from the middle, so that it cannot be drawn asunder either lengthwise or sidewise.

TAFFAREL, OR TAFF-RAIL.—The upper part of the ship's stern, usually ornamented with carved work or mouldings, the ends of which unite to the quarter-pieces.

TASTING OF PLANK OR TIMBER.—Chipping it with an adze, or boring it with a small auger, for the purpose of ascertaining its quality or defects.

TO TEACH.—A term applied to the direction that any line, &c., seems to point out.— Thus we say "let the line or mould *teach fair* to such a spot, rase," &c.

TENON.—The square part at the end of one piece of timber, diminished so as to fix in a hole of another piece, called a mortise, for joining or fastening the two pieces together.

THICKSTUFF.—A name for sided timber, exceeding four inches, but not being more than twelve inches, in thickness.

THROAT.—The inside of knee timber at the middle or turn of the arms. Also the midship part of the floor timbers.

TOP AND BUTT.—A method of working English plank so as to make good conversion. As the plank runs very narrow at the top clear of sap, this is done by disposing the top end of every plank within six feet of the butt end of the plank above or below it, letting every plank work, as broad as it will hold clear of sap, by which method only can every other seam produce a fair edge.

TOPSIDE.—A name given to all that part of a ship's side above the main-wales.

TOP-TIMBERS.—The timbers which form the topside. The first general tier which reach the top are called the long top-timbers, and those below are called the short top-timbers. *See* FRAMES.

TOP TIMBER LINE.—The curve limiting the height of the sheer at the given breadth of the top-timbers.

TOUCH.—The broadest part of a plank worked top and butt, which place is six feet from the butt end. Or, the middle of a plank worked anchor-stock fashion. Also the sudden angles of the stern-timbers at the counters, &c.

TRAIL-BOARDS.—A term for the carved work, between the cheeks, at the heel of the figure.

TRANSOMS.—The thwartship timbers which are bolted to the stern-post in order to form the buttock; and of which the curves, forming the *round aft*, are represented on the horizontal, or half-breadth plan of the ship.

TREAD OF THE KEEL.—The whole length of the keel upon a straight line.

TREENAILS.—Cylindrical oak pins driven through the planks and timbers of a vessel, to fasten or connect them together. These certainly make the best fastening when driven quite through, and caulked or wedged inside. They should be made of the very best oak, cut near the butt, and perfectly dry or well seasoned.

THE TUCK.—The aft part of the ship where the ends of the planks of the bottom are terminated by the tuck-rail, and all below the wing-transom when it partakes of the figure of the wing-transom as far as the Fashion-pieces. *See* SQUARE TUCK.

TUCK-RAIL.—The rail which is wrought well with the upper side of the wing-transom, and forms a rabbet for the purpose of caulking the butt ends of the planks of the bottom.

WALL-SIDED.—A term applied to the topsides of the ship when the main breadth is continued very low down and very high up, so that the topsides appear strait and upright like a wall.

WASH-BOARD.—A shifting strake along the topsides of a small vessel, used occasionally to keep out the sea.

WATER LINES, OR LINES OF FLOATATION.—Those horizontal lines, supposed to be described by the surface of the water on the bottom of a ship, and which are exhibited at certain depths upon the sheer-draught. Of these, the most particular are those denominated the *Light Water Line* and the *Load Water Line*; the former, namely, the Light Water Line, being that line which shews the depression of the ship's body in the water, when light or unladen, as when first launched; and the latter, which exhibits the same when laden with her guns and ballast, or cargo. In the Half-Breadth Plan these lines are curves limiting the Half-breadth of the ship at the height of the corresponding lines in the Sheer Plan.

WATER WAYS.—The edge of the deck next the timbers, which is wrought thicker than the rest of the deck, and so hollowed to the thickness of the deck as to form a gutter or channel for the water to run through to the scuppers.

WHOLE MOULDED.—A term applied to the bodies of those ships which are so constructed that one Mould made to the midship bend, with the addition of a floor hollow, will mould all the timbers, below the main breadth, in the square body.

WINGS.—The places next the side upon the orlop, usually parted off in ships of war, that the carpenter and his crew may have access round the ship, in time of action, to plug up shot-holes, &c.

WING-TRANSOM.—The uppermost transom, in the stern frame, upon which the heels of the counter timbers are let in and rest. It is by some called the main transom.

WOOD-LOCK.—A piece of elm, or oak, closely fitted, and sheathed with copper, in the throating or score of the pintle, near the load water line; so that, when the rudder is hung, and the wood-lock nailed in its place, it cannot rise, because the latter butts against the under side of the Brace and butt of the score.

WRAIN BOLTS.—Ring bolts, used when planking, with two or more forelock holes in the end for taking-in the sett, as the plank, &c., works nearer the timbers.

WRAIN STAVE.—A sort of stout billet of tough wood, tapered at the ends so as to go into the ring of the wrain bolt to make the setts necessary for bringing-to the planks or thick-stuff to the timbers.

ERRATA.

Page 20,	top of the page,	for "heany,"	read	heavy.
" 54,	" " " "	" "qualities,"	"	quality.
" 74,	3d line from side,	" "qualities,"	"	quantities.
" 77,	last line,	" "heigth,"	"	height.
" 86,	13th line from top,	" "mnst,"	"	must.
" 99,	last line,	" "square,"	"	squares.
" 136,	39th, 40th, 41st, and 42d lines,	" "ordinate,"	"	ordinates.
" 136½,	last line,	" "No. II,"	"	No. XI.
" 138,	33rd line,	" "movements,"	"	moments.

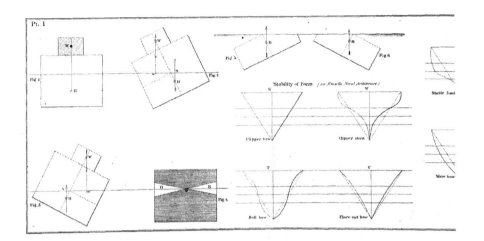

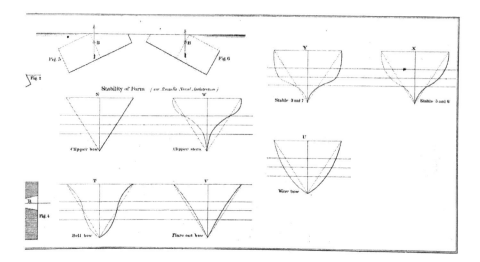

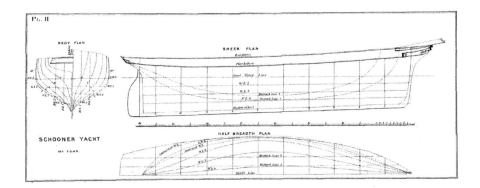

Pl. III

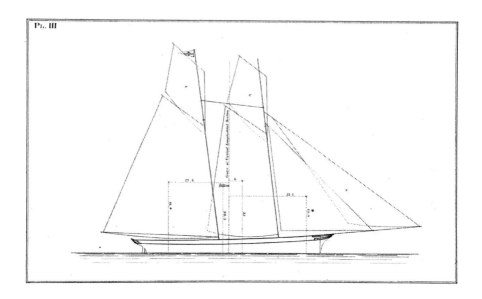

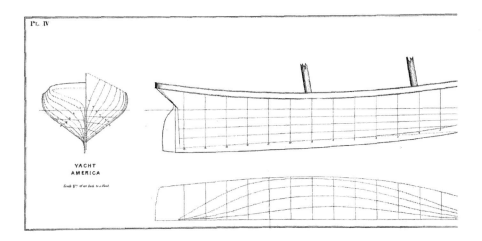

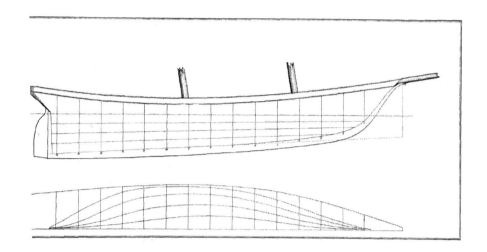

CPSIA information can be obtained
at www.ICGtesting.com
Printed in the USA
LVHW081643040122
707839LV00002B/50